A TEXT BOOK

OF

CHEMICAL ENGINEERING

A TEXT BOOK

OF

CHEMICAL ENGINEERING

BY

EDWARD HART

GYAN

Published by

Gyan Publishing House
5, Ansari Road
Daryaganj, New Delhi-110002
Phone: 011-47034999, 9811692060
E-mail: books@gyanbooks.com

Distribution Network
gyanbooks.com
India, USA, Canada, UK, Australia

ISBN: 978-93-6280-223-1 (HB)
First Published, 1920

2nd Impression 2024

Printed at: Gyan Press, Delhi.

A TEXT BOOK OF CHEMICAL ENGINEERING
Author: EDWARD HART

A TEXT BOOK

OF

CHEMICAL ENGINEERING

BY

EDWARD HART, Ph. D.

Professor of Chemical Engineering in Lafayette College. Sometime Editor of The Journal of Analytical and Applied Chemistry, and of The Journal of The American Chemical Society, President of The Baker & Adamson Chemical Co., and Vice-President and General Manager of The Clinchfield Products Co.

EASTON, PA.
THE CHEMICAL PUBLISHING CO.
1920

LONDON, ENGLAND:
WILLIAMS & NORGATE,
14 HENRIETTA STREET, COVENT GARDEN, W. C.

TOKYO, JAPAN:
MARUZEN COMPANY LTD.,
11-16 NIHONBASHI TORI-SANCHOME

PREFACE.

For several years I have accumulated data with the intention of eventually publishing a book on chemical engineering to embody the elements of the subject. Publication has been hastened by the pressing need of a text book for my own classes. I realize more fully than any of my critics can do the imperfections of omission, and I fear some of commission in this first attempt to give a comprehensive view of the subject. The writer whose own experience shall fully qualify him to cover this whole subject is still to be discovered. I can only hope that this imperfect presentation may be allowed to pass muster as a first attempt. I am greatly indebted to many persons, some of whose names are mentioned in the text, for assistance. I am also indebted to many manufacturers for cuts of apparatus, for special drawings and other favors. The names of such firms inserted in the text have had the purpose of helping the beginner. They were not paid for in any way. If future editions should be called for I hope with the assistance which criticism of this one shall evoke to prepare a much better treatise on the subject. For all such criticism I shall be very grateful, especially if it be kindly criticism.

EDWARD HART.

Easton, Pa., Dec. 1, 1920.

Contents.

List of Illustrations

Chemical Engineering

INTRODUCTION.

(1) The field of chemical engineering is a very large one. To cover it properly requires the combined experience of many men. It is bound up with chemical technology and to be of real use, statements must be not only scientifically precise but keep in mind constantly the question of income and expense. Where works are already in operation it is not unusual to find that the practice has grown antiquated, but the young engineer should go at the task of improving cautiously. Those deriving a good income from existing operations are slow to see the necessity of spending money to improve the product. They are shy of too much theory, especially when propounded by young engineers. The beginner is urged to improve quality and output with existing apparatus and so inspire confidence before suggesting changes.

Improved processes should always be tried out in the laboratory, then on a small manufacturing scale before being introduced as a part of the large scale operation. Neglect of this precaution has led to serious loss.

CHAPTER I.

MATERIALS.

It is a well established fact that we can make anything if we have something to make it in. For small scale operations this causes little difficulty but often becomes serious when working on ton lots.

(2) **Iron**[1] is usually the cheapest material for containers. Means for casting all but very large pieces are available locally, thus enabling careful superintendence while casting and avoidance of mistakes and heavy transportation charges.

Cast iron may be used as a container for sulphuric acid of over 60 per cent. strength at ordinary temperatures, or for 70 per cent.

[1] See also Fawsitt & Powell, *J. S. C. I.*, 1914, 234 and Fawsitt *Ib.*, 1920, 147T.

acid at 100° C.; the higher the temperature the more iron will dissolve. Cast iron resists fuming sulphuric acid better than mild steel, but alternate contact with acid and air develops cracks.

(3) Mild Steel of say 0.15 per cent. carbon such as is used for tanks, boilers, etc., is acted upon by sulphuric acid to a greater extent than cast iron. It is the best material for storing oleum, but prolonged storage of oleum containing from 100-105 per cent. H_2SO_4 (0-20 per cent. SO_3) should be avoided as it attacks steel much more rapidly than when above or below this limit.[1] Mild steel is also best suited for mixed nitric and sulphuric acids and may be used for all mixtures containing not less than 12 per cent. sulphuric acid with a total acidity of 90 per cent. at ordinary temperature. As the temperature rises the action on the steel increases, and as this action is exothermic, it may easily become dangerous. It may be bent into shape and melted together by the oxyacetylene blowpipe. Unprotected cast iron or steel can not be used as a container for ferric solutions, chromic acid or solutions of copper, mercury or silver. Solutions of the alkalies do not affect them.

HIGH SILICON IRON ALLOYS

such as Duriron and Tantiron are resistant to dilute sulphuric and nitric acids. They suffer, however, from lack of strength, being easily cracked by a blow or by too rapid heating.

(4) Composition.—"Duriron, which is typical of these alloys and was furthermore the first in the field in this country, is referred to throughout this article. Its approximate analysis follows:

		Per cent.
Silicon	about	14.00
Carbon	below	0.60
Manganese	"	0.50
Sulphur	"	0.10
Phosphorous	"	0.20
Iron	about	85.00

In a general way, other things being equal, the rate of depreciation will vary inversely with the silicon content, between the limits of 10 and 25 per cent."

[1] See Knietsch, *Ber.* 34, 4108, 1901.

"This does not, however, hold true for the lower ranges, as experiments show that alloys containing between 7 and 8 per cent. of silicon are much more readily attacked than is the case with grey iron having only 3 to 4 per cent., and furthermore white irons, with 1 per cent. or less, resist weak sulphuric acid much better than where the content is from 1-10 per cent."

"While alloys of 18 to 25 per cent. of silicon, or thereabouts, are so resistant as to withstand even hot hydrochloric acid—the presence of such quantities render them so brittle that they are practically useless for commercial purposes."

"An excessive manganese content, on the other hand, say from 1 to 2 per cent., adds to the strength of the casting and softens it to the extent of making it machinable, but has the drawback of lowering the acid resisting properties of the alloy."

(5) Physical Properties.—"Duriron is a cast metal extremely hard and close grained. It shows a white fracture, taking and retaining a better polish than nickel."

"It has a specific gravity of 7.00—is not as strong as cast iron—has a coefficient of expansion two and one-half to three times greater than that of cast iron or steel—a heat conductivity about that of cast iron and an electrical conductivity only about one-fortieth that of standard annealed copper. Contraction allowance for casting is 3/16 inch per foot."

"Duriron will not soften materially, nor lose its shape, at a temperature below its melting point, about 2500° F. It is very fluid in the molten state, permitting pouring into very thin sections when desirable, and shows practically no oxidation at even the highest temperatures."

"Owing to its extreme hardness it is highly resistant to *erosion* and cannot be machined with a cutting tool, but must be finished by grinding."

(6) Chemical Properties.—"Duriron is practically unaffected by any strength of nitric, sulphuric, acetic and most other acids or alkalies. The chief exception is commercial concentrated hydrochloric, which causes marked action; this lessens as the acid is diluted, until with 1 or 2 per cent. there is no loss even when boiling. Samples of the alloy exposed continuously for five weeks to hydrochloric of different strengths, at temperatures ranging between 178° to 185° F. lost weight as follows:

In 25 per cent. acid (by weight).............. 1.72 per cent.
In 10 per cent. acid (by weight).............. 0.74 per cent.

with hydrochloric acid gas the resistivity is highly satisfactory."

"Phosphoric of certain strengths can be handled, others attack readily. Strong aqua regia and hydrofluoric cannot of course be withstood."

"The following laboratory tests, at room temperature, are interesting, as illustrating comparative resistivity.

In 25 per cent. sulphuric (by weight):
Duriron..........................No loss in one year
Cast iron................Completely dissolved in 16 days
Wrought iron...........Completely dissolved in 30 days
In 25 per cent. nitric (by weight):
Duriron...........In one year lost 11/1000 of 1 per cent.
Cast iron...............Completely dissolved in 63 hours
Wrought iron..........Completely dissolved in 58 hours.

As a general rule Duriron is not suitable for fusions, though there are some exceptions."

(7) Design of Apparatus.—"Owing to the high coefficient of expansion and the various manufacturing difficulties encountered, careful attention to a number of points are necessary in designing apparatus that can be economically and successfully produced in this material."

"While great progress has been made in producing large and difficult castings, it is important to avoid extensive flat surfaces, abrupt changes in section and sharp angles."

"Curved surfaces are not only stronger than flat, but they are better able to resist temperature changes without cracking."

"It is not good practice to use chaplets in supporting cores, consequently all coring should be made as simple as possible."

"As the alloy is too hard to be economically drilled, holes should be cored."

"Hemispherical bottoms for kettles are best and should be used wherever possible."

"While covers for kettles, and similar apparatus, can be made almost flat, if not too large, it is far better practice to give them considerable arch, three or four inches per foot of diameter not being excessive. Openings in covers should not be placed too close together or in direct line across the cover."

(8) Installation of Apparatus.—"While Duriron is more easily broken than cast iron, it will be found amply strong if properly handled and installed."

"Avoid unnecessary strain on every part."

"Line up and support all members carefully before tightening connections."

"Pipe lines should be anchored, especially at bends, and where-ever attached to flanged apparatus."

"As expansion of Duriron is about two and a half to three times that of iron or steel, allowance must be made for this when installing."

"Draw up equally on every bolt and use only enough pressure to prevent leakage. If gasket is soft enough a 6-inch wrench held between thumb and forefinger will be sufficient."

"Hard gaskets must not be used. They should be fairly soft and preferably at least 1/16 inch thick. The selection of material for gaskets and packing should be based on kind of acid handled."

"Do not use hard cements for bell joints. These prevent expansion and may cause cracking."

(9) Forms in Which Furnished.—"The alloy is furnished in the form of cocks, valves, pumps, pipe and fittings, steam syphons, exhaust fans, jacketed and plain kettles, stills, towers, condensers, concentrators, in fact in any shape from a few ounces up to five tons or more, of a design suitable to the characteristics of the material."[1]

(10) Copper is largely used for stills, separating columns and vacuum pans. It is not suitable for use with most acid liquids nor ammonia.

(11) Lead and Lead Lined vessels are very largely used in chemical apparatus as the metal is indifferent or nearly so to many chemical solutions. Many leaden articles such as pumps and fans are cast of hard lead, and an alloy of lead and antimony,

[1] This article on the properties of the iron silicon alloys I owe to the kindness of Mr. J. R. Pitman of the Duriron Co.

which is stiffer and harder than pure lead. Lead burning is an art by itself and large chemical works nearly always have several lead burners. Lead is soft, and when heated expands but has not sufficient tensile strength to resume its original form when cooled. As a result lead linings often pucker or become covered with ridges and are rapidly destroyed. Lead lined pipe shown in Figs. 1 and 2 is cheaper and stiffer than lead pipe. Ordinary lead pipe

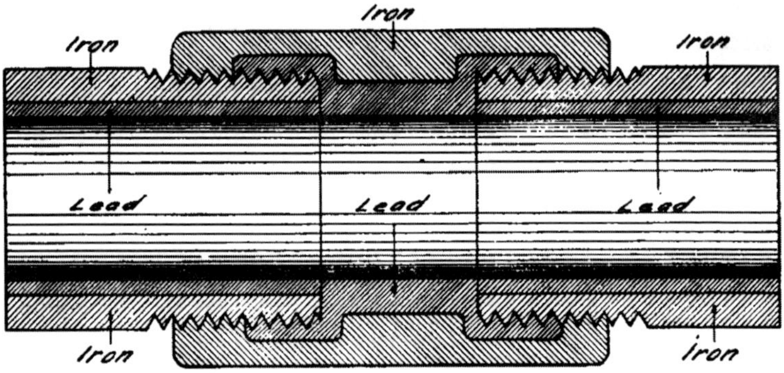

Lead Lined Iron Pipe

Fig. 1.

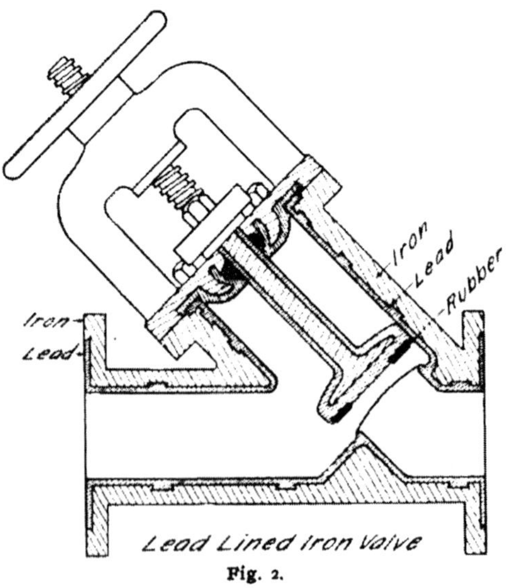

Lead Lined Iron Valve

Fig. 2.

of 1 inch diameter will sustain a pressure of say 60 pounds and may be used in evaporators. Lead may be used for all strengths of sulphuric acid up to 80 per cent. at all temperatures up to the boiling point. Above 80 per cent. it is attacked rapidly at high temperatures but not at ordinary air temperatures up to 98 per cent., over 98 per cent. it is rapidly attacked. Acetic acid in presence of air dissolves lead rapidly. In a vacuum the oxide on the surface of the lead is removed, but the lead itself is not affected.

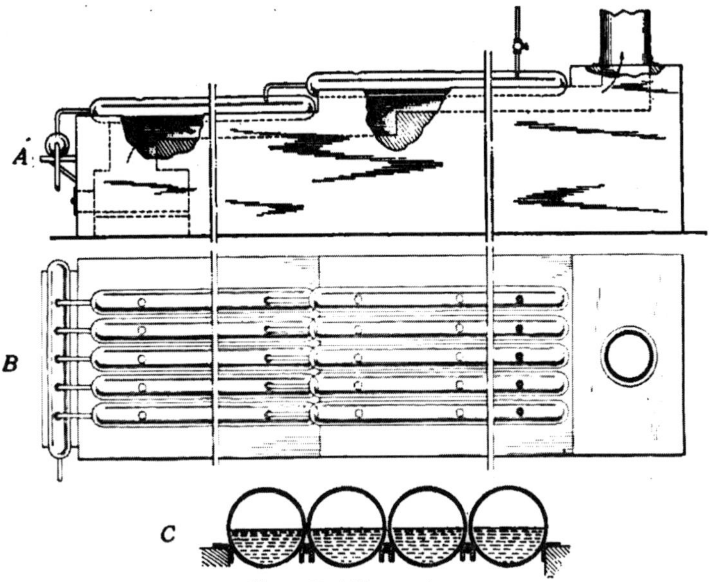

Fig. 3. Hart Evaporator.

(12) Aluminum, cast or spun, can now be had in large sizes. It dissolves very slowly in nitric, sulphuric and acetic acids but rapidly in hydrochloric acid and some chlorides and in alkalies.

(13) Glass vessels are not available for large size apparatus. Glass tubing up to 3 inch diameter and test tubes up to 1½x18 inches may be had and these are used in the Hart boilers (Fig. 3) and condensers (U. S. Patents 603,508; 1,096,838; 1,254,689; 525,761 and 601,466). For larger apparatus the Pfaudler Co., of Rochester, N. Y., supplies glass-lined apparatus shown in Figs. 4-8. The Elyria Enameled Products Co., Elyria, Ohio, also sup-

Fig. 4.—Jacketed Open Mixing Tank.

Fig. 5.—Open Mixing Tank and Evaporator.

Fig. 6.—One-piece Jacketed Closed Mixing Tank.

Fig. 7.—Jacketed Closed Mixing Tank.

Fig. 8.—Glass Lined Ware made by the Pfaudler Co.

ply this ware upon a cast-iron base. This company states that their glass enamel has the resisting qualities of chemical glass. It is fused into the metal itself and will not crack, scale or craze. Fig. 9 shows three sizes of vacuum still, and Fig. 10 a condenser made by this company. Stuart and Peterson, Burlington, N. J., supply enameled cast-iron ware for similar purposes.

Fig. 9.—Three sizes of Enameled Vacuum Still.

(14) Fused silica is now being used to a considerable extent in the manufacture of chemical ware, taking the place of glass. It is equally fragile but endures sudden changes of temperature much better than glass. It is furnished by the Thermal Syndicate, Ltd., 50 E. 41st St., New York City, and Chas. Engelhard, 30 Church St., New York City.

(15) **Chemical stoneware** is supplied in great variety by the General Ceramics Co., 50 Church St., New York City, Chas. Graham, 986 Metropolitan Ave., Brooklyn, N. Y., U. S. Stoneware Co., Akron, Ohio, and Maurice A. Knight, E. Akron, Ohio.

(16) **Silicon** has been proposed as a material for chemical ware, but has not been largely used.

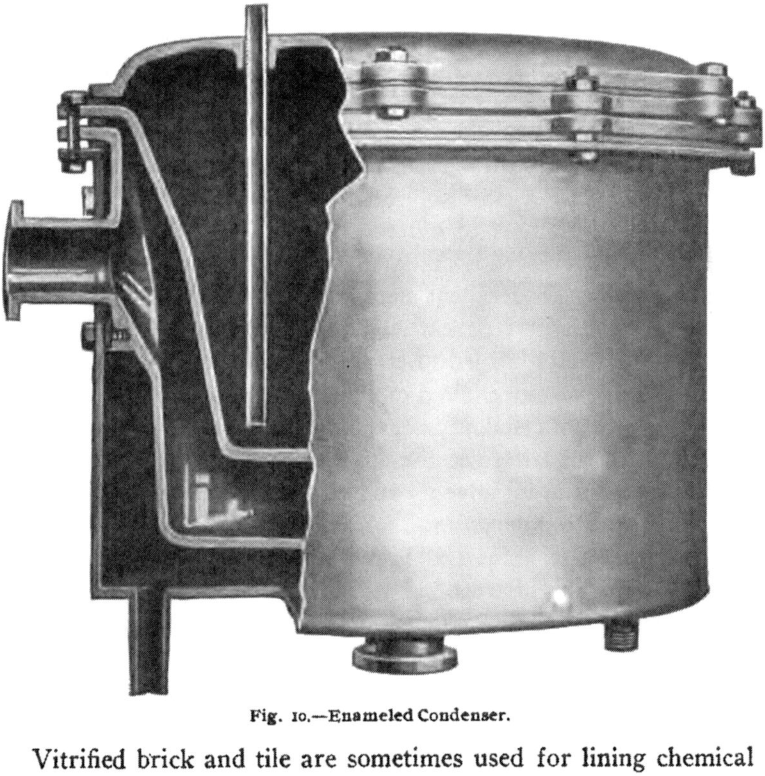

Fig. 10.—Enameled Condenser.

Vitrified brick and tile are sometimes used for lining chemical containers. These may be obtained from the above named manufacturers of chemical stoneware or from the Harbison Walker Co. of Pittsburgh, Pa.

(17) "**Bakelite**"[1] is obtained by heating phenol or cresol with formaldehyde preferably in presence of ammonia or other alka-

[1] This description of the properties of Bakelite was kindly contributed by Dr. Baekeland himself.

line condensing agents. In view of the fact that ammonia in presence of formaldehyde is immediately transformed into hexamethylenetetramine it is possible to use directly the latter chemical in part or in totality, instead of the usual commercial formaldehyde solution. Other substances which are the equivalent of formaldehyde as for instance its polymer paraform, or methylal may be used for the same purpose.

The resulting phenol condensation product is hard and transparent and has the appearance of amber; but its chemical and its other physical properties are decidedly different. For instance, if heated it does not melt and if carried at higher temperatures it chars and blackens without however liquifying. Furthermore it is insoluble in all organic solvents and withstands most chemicals which act upon amber or other resins. It is harder than amber and its tensile strength is much higher. It has some analogy with amber by being an excellent electrical insulator and this property in conjunction with its great strength as well as its chemical and physical resistivity has led to endless technical applications where resinous substances, celluloid or rubber are unsuitable.

It differs from celluloid in that it lacks the great flexibility and toughness of the latter; on the other hand it is harder, withstands higher temperatures, solvents and chemicals and offers a greater scope for molding purposes.

The general technique of bakelite consists in interrupting the chemical reaction between phenol and formaldehyde at a definite phase where an initial condensation product is obtained which is called A.

This initial product A is still fusible and soluble and possesses some of the same general properties of the ordinary resins.

It can be kept in this condition indefinitely. However, if it be submitted to further heating it will melt at first but on further application of heat, the molten mass will thicken and change into a hard infusible insoluble substance which is the final product C, much harder and stronger than the product A from which it is derived through a process of polymerization. This latter change is exothermic; by reason of this, gaseous products form in the mass while it hardens and tend to produce objectionable porosity.

In order to prevent this defect, the heating is carried out under counter pressure either in closed molds or in a tank with gas or oil under pressure.

Porosity occurs especially whenever hexamethylenetetramine or much ammonia is used in the reaction which liberates ammonia gas in the final stage. On the contrary when a fixed alkali, for instance caustic soda, is substituted, then the polymerization proceeds with much less danger of porosity and counter pressure can be omitted.

Under certain well established conditions, for instance when the proportion of phenol is relatively high and when the reaction proceeds in an acid medium, totally different resinous condensation products are obtained which resemble more the true natural resins in that they are permanently fusible and soluble. They are known under various names: Novolak, saliretin, phenolic resin, fusible phenolic resin, etc.

Although they are derived from the same original products, phenol and formaldehyde, and notwithstanding the fact that at first sight they may resemble bakelite A, their differentiation from the latter product is very simple.

If they are heated, they simply melt like any natural resin and they remain liquid as long as the temperature is kept up; on cooling, the liquid solidifies and the resulting resinous mass is brittle and can be melted again; nor has it lost its solubility. Under the same treatment, bakelite A, on the contrary, if submitted to heat will melt, solidify *without cooling* into an infusible insoluble product much harder and stronger than the Novolak resins.

Means have been found, however, to chemically transform Novolak into Bakelite C by heating it with paraform, or with formaldehyde and ammonia or with hexamethylenetetramine."

It has a specific gravity of 1.25, and is indifferent to all solutions except strong sulphuric and nitric acids. It is a very poor conductor of heat and electricity. It is readily worked in the lathe and may be sawed and filed.

(18) Celluloid is a plastic made by incorporating nitrocellulose with camphor with the aid of ethyl alcohol. It softens when heated and may be readily molded into articles which are flexible and tough. It is somewhat unstable. At ordinary temperatures it slowly decomposes and at higher temperatures this takes place more readily. The camphor slowly volatilizes and by the action of alkaline solutions the nitrocellulose is decomposed. For chemical use it is probably most commonly applied as a varnish.

A large variety of paints and varnishes have been used as protective coatings for chemical apparatus. Among these are nitrocellulose varnishes, manufactured by Du Pont de Nemours Co., Wilmington, Del., and the R. I. W. paint of Toch Bros., 320 5th Ave., New York City.

CHAPTER II.

LOCATION OF WORKS.

(19) Where a new plant is to be built the location is an important question. Many items enter into this problem. In the first place, it is desirable to have the plant located near a city of some size, so that minor supplies may be quickly obtained. This is also desirable because it is then easier to obtain and keep labor, and the heavy initial cost of company dwellings, clubs, etc., need not be met.

Where materials or articles manufactured come or go over seas, the factory should, if possible, be located on tide water so that vessels may anchor alongside the works. The same holds if they are best transported along the coast.

The plant should be so located that the raw materials may be obtained cheaply with low freight rates. In studying this part of the problem, all supplies including coal, water, gas, electricity, and most important of all, labor must be considered. Location on competing railroads is desirable, but can be obtained only in exceptional cases. Low cost of the land is a consideration which may sometimes be disregarded because of greater proximity to street car lines or other means of communication. One of the important considerations is proximity to markets.

After the location has been fixed, a topographical and profile survey should be made in order that the plans may be laid down directly on the blue prints.

(20) Laying Out of Plant.—It is desirable when possible to have the heavy substances unaffected by weather shipped in coal gondola cars, delivered on trestles, and dumped through the bottom. Two styles of car are made for this purpose. One with bottoms slanted the entire length and the other delivering only a part of the content, the remainder having to be shoveled out. The first are to be preferred when obtainable. Coal cars generally run light one way, and railroad companies will sometimes make a reduced rate when return freight is provided. Box cars require more labor in unloading.

The cheapest way to handle heavy low priced material is to locate the plant on a hillside and so arrange it that the crude material is delivered at the highest level, gradually passing to a lower as the result of each successive operation, until it reaches the warehouse whose floor should be on the level of the car floor.

Where the lay of the land permits, ore pockets may be provided under trestles. They are best made with sloping floor so as to deliver easily. They must be of such size as to hold easily not less than one carload. In low tracts of land where floods may be feared, the foundation walls of buildings are sometimes raised four feet above the surface and filled to raise the floor to that level. This brings the car floor on a level with the building floor, the car tracks being laid in wharves between the building 16 feet wide. Since the car is commonly only 8 feet wide, this seems like a waste of space. Four feet space on each side is necessary, however, for the safety of the railway employees in coupling and uncoupling. This space must not be encroached upon by stairways or other obstructions. Stairways are best built in the shipping platforms outside the buildings. Shipping platforms are best made of concrete; they should not be less than 9 feet wide. When it is necessary to provide wharves, the building should be carefully spaced so that when a second row of buildings is built, the ends may coincide, or nearly coincide, with those of the first row. If this is not looked after, it will be impossible to arrange track crossings for teams between the buildings. The shipping platforms should be covered from the weather by allowing the roofs to project over them. So far as necessary these projections may be supported by brackets fastened to the walls.

Not less than twenty feet space should be allowed between the ends of the buildings. If there are three or more rows of buildings, an occasional space of 60 feet should be allowed. This gives room for turning teams, for mixing concrete for repairs, etc.

On level ground reasonably safe from floods, the floors and railway tracks must be laid on one level and elevators and conveyors provided for transferring material from one place to another.

Railway Sidings.—Minimum curves for standard tracks are 16 degrees for sidings and 3 degrees for main lines. Where the line is double tracked, facing point switches are not allowed.

(21) Water Supply.—In many operations the amount and quality of the water supplied is a matter of great importance. The water available should be carefully examined and the seasonal variations of supply and quality ascertained. Springs that dry up or are so located as to be flooded and supply only dirty water during part of the time are not satisfactory. Limy water or water loaded with organic matter may be altogether unsuitable. It is very desirable, where possible, to have an independent supply of water so located that pumping is not necessary.

(22) Disposal of Waste.—The disposal of waste is another serious problem. Where noxious vapors like hydrogen sulfide are likely to be liberated from the waste by accident or carelessness, works should never be located near houses nor so located that waste water will flow through thickly inhabited districts.

The disposal of solid waste is often a troublesome problem. Near or on tide water this as well as liquid waste may often be run into the ocean. Niter cake, residues from barium sulphide, calcium chloride mud with lime, and various other materials each offers a separate problem.

BUILDINGS.

The material to be used for buildings will naturally be largely determined by local conditions. In calculating, the price and freight rates are most important.

(23) Concrete.—A good cheap mixer for concrete is that made by the Waterloo Cement Machinery Corporation, of Waterloo, Iowa (Fig. 11). One form of this apparatus receives the materials in a scoop at the ground level, elevates and discharges into the mixer and when mixing is complete, it discharges into wheelbarrows. One of these mixers travels under its own power, climbing 20 per cent. grades; it is operated by an 8 horse-power gasoline engine. These mixers are built in sizes from 5 to 11 cubic feet per batch. Another mixer which gives good service is built by the Jaeger Machinery Co., of Columbus, Ohio. A third

Fig. 11.

mixer is made by the Municipal Engineering Co., of Chicago, Ill., etc.

(24) The Cement Gun.—The Cement Gun Co., of Allentown, Pa., claim that in order to get maximum strength the cement material must be applied at once on wetting so as to avoid disturbance after initial set begins. For this reason the mixed material is placed in a hopper and forced through a pipe by compressed air. At the moment of discharge it is properly wetted by water supplied through a special nozzle and auxiliary hose. The muzzle velocity of the material corresponds to a pressure head of 35 pounds (Fig. 12). This process is evidently specially useful for distributing cheaply thin layers of material on ceilings, walls, for building tanks, etc. It is also claimed that the material has a smaller percentage of voids and is therefore less permeable, that it has greater tensile strength and adheres better than the hand-made product.

(25) Brick.—Everything considered, brick is the best material for chemical buildings, as it is very slightly attacked by acid gases, is fireproof and usually costs little more than wood. For muriatic or nitric acid houses, it should be laid in cement. Where cranes are to be installed the walls may be strengthened by placing I beams vertically at intervals of, say 20 feet. The steel roof girders will then be riveted to the upper end. This construction makes a very attractive building but the ironwork must be carefully protected from acid fumes, by means of one or more coats of asphalt varnish. This must be watched and promptly renewed as soon as it shows signs of peeling off.

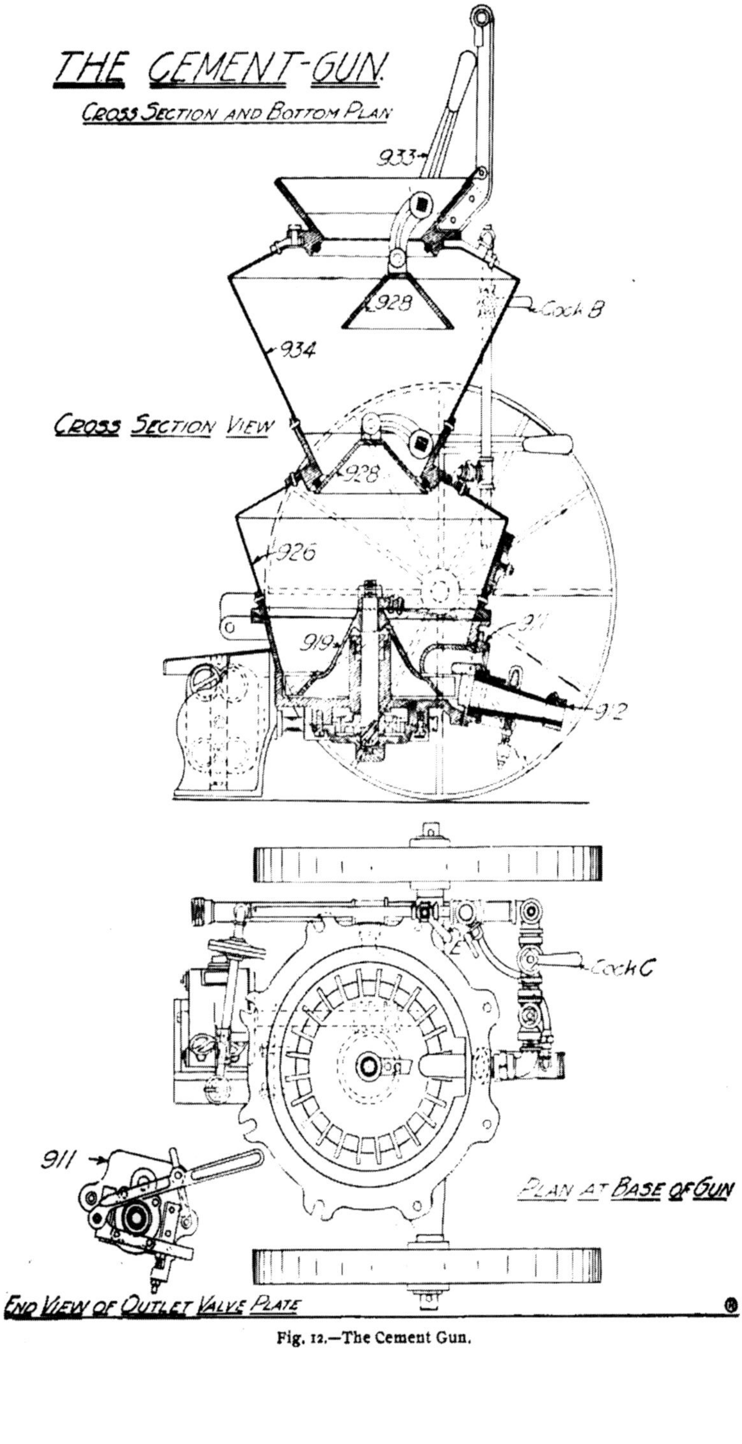

Fig. 12.—The Cement Gun.

FIG. 12—THE CEMENT GUN.

The Cement Gun consists essentially of a hopper into which the cementitious materials are placed; a hose connected to the bottom of the hopper, through which the dry mixture is forced by compressed air; and a specially constructed nozzle at the end of the hose to which is attached a second hose for the purpose of supplying water for hydrating the material.

HOW THE CEMENT GUN IS OPERATED.

Before admitting compressed air to tanks, 926 and 934, cone valves 928 are open resting against sides of the cylinders, leaving a clear opening for material to be charged into both tanks. After dry material (mixed in the proper proportions) is placed in tanks, upper cone valve 928 is seated by means of the lever 933 and compressed air is admitted through cock "B," thus holding cone valve 928 in seat. Cock "C" is then opened admitting air through the gooseneck 911, and outlet valve 912 (cock "E") is then opened which makes feed-wheel 919 revolve, and the material in each pocket of same, as it registers with gooseneck 911, is blown out through valve 912 into the material hose attached thereto, and thence to the nozzle *where the water from a separate hose is added.*

To recharge machine, it is not necessary to stop, as lower cone valve 928 is seated and held in position shown, when air in upper tank is exhausted to atmosphere. Upper cone valve 928 then opens and a new charge is put into the machine.

(26) Hollow Tile have been much used recently in construction of chemical buildings. They make good buildings, are reasonable in price, and resist the attack of acids. Fig. 13.

(27) Framed Buildings of wood are much cheaper in many parts of the country. When used for acid houses they must be framed together; nailing will not answer. The timbers are to be mortised and tenoned and fastened with hickory pins. Siding is nailed on, but where much acid escapes, the nailing must be renewed from time to time. Zinc nails manufactured by the N. J. Zinc Co., are said to last well. One of the many composition roofs now on the market seems to answer best for chemical buildings. Slate usually contains some calcium carbonate, and does not last. Flat slag and tar roofs are cheap and last well.

Good paving brick laid in sand, make about the best floors, but are expensive. Concrete floors will not stand acid. Concrete protected by a coating of boiled tar lasts well. The tar must be put on after the cement has been thoroughly dried.

(28) Foundations.—Much depends upon securing good foundations, but reliable data are nearly always hard to obtain. It is

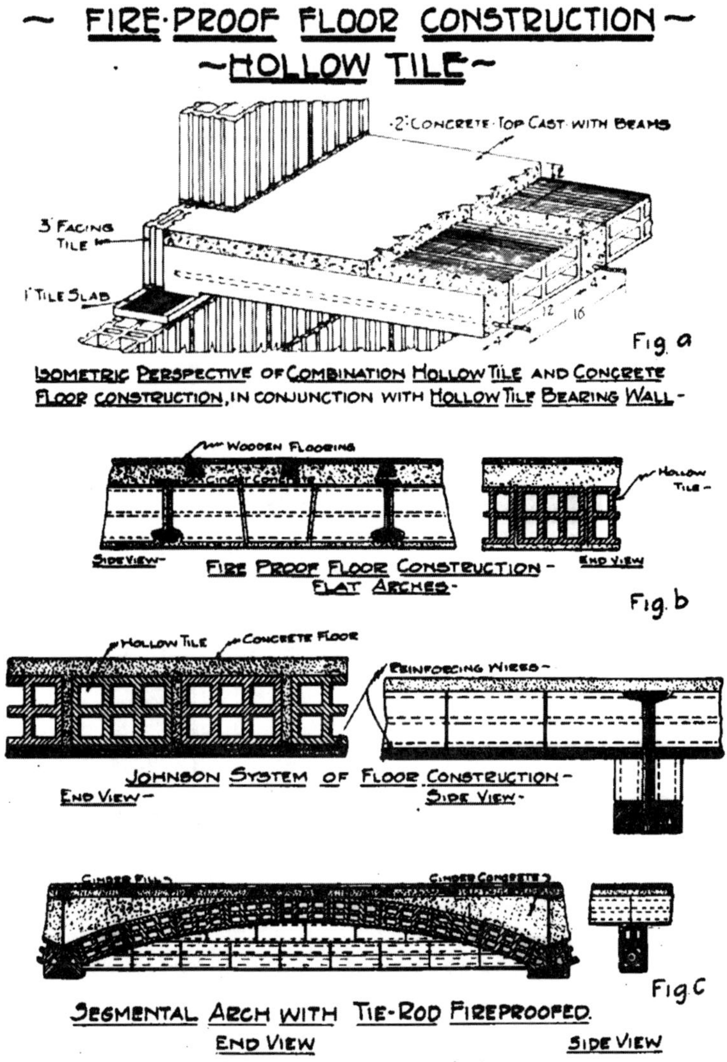

Fig. 13.—Hollow Tile Construction.

impossible, in most cases to determine with accuracy the resistance to deformation of the foundation and hence, this must be left to the judgment of the engineer. In a few cases, where heavy weights must be carried the bearing capacity must be determined. A description of the method used with the New York State Capitol at Albany will show how this may be done: A hole 3 feet deep was dug into the blue clay, 18 inches square at the top and 14 inches square at the bottom. Rows of small stakes were then driven in lines radiating from the center of the hole and the tops of the stakes carefully leveled. A timber post 12 inches square held upright by guys and with cross-staging on the top to hold the weights was placed in the holes. The level of the tops of the stakes remained unchanged until a load of 5.9 tons per square inch was applied.

(29) Safe Bearing Loads.—These are about as follows:

Hard solid rock	200 tons per square foot
Ordinary rock	20–30 tons per square foot
Hard gravel	6–10 tons per square foot
Sandy gravel	4–6 tons per square foot
Clay	3,500–4,500 tons per square foot

The New York building law allows for:

Soft clay	1 ton per square foot
Ordinary clay	2 tons per square foot
Loam, clay or fine sand	3 tons per square foot
Coarse sand, stiff gravel or hard clay	4 tons per square foot

A foundation should always be horizontal or stepped (Fig. 14),

Fig. 14.—Stepped Foundation.

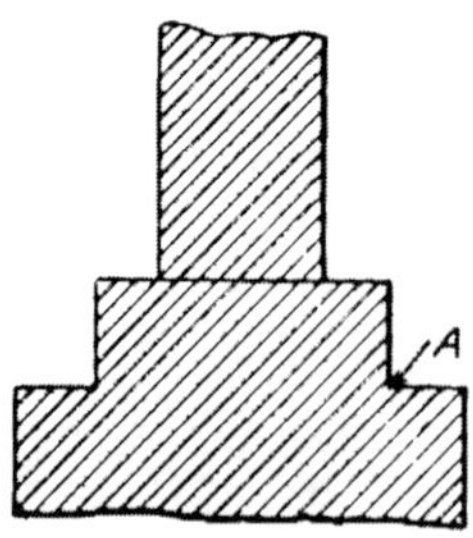

Fig. 15.—Foundation Footing.

BRICK LAYING.

Fig. 16.
Improper Method—daubing the corner.

Fig. 17.
Proper Method—shoving up the brick.

and where practicable should resist pressure equally throughout. Thus, it is undesirable to rest one side of a building upon solid rock and another upon hard clay, though this is sometimes unavoidable.

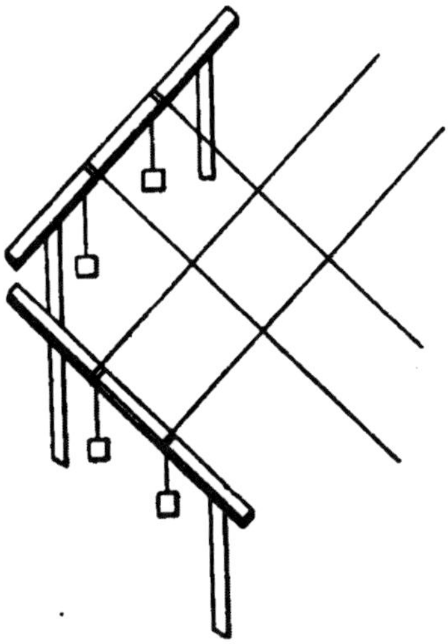

Fig. 18.- Corner Layout.

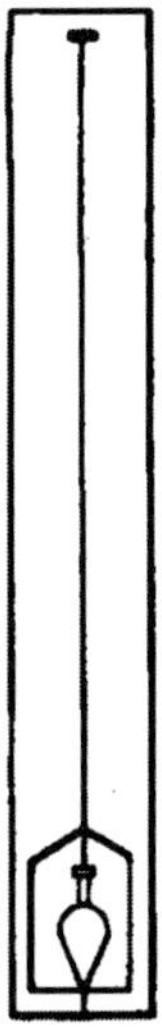

Fig. 19.—Plumb Board.

(30) Footings.—These are used to extend the area subject to compression and are especially desirable where the foundation is of doubtful rigidity. By giving the wall a greater width over the bottom surface, the weight upon unit area is decreased and likewise the danger of settling. In Fig. 15 the bottom course, usually of cement concrete, should be made so thick that no danger of fracture exists at A. The cement mass may be and often is re-enforced. It is always necessary that trenches for foundations reach below the frost line; in the latitude of New York to, say, 3 feet below the surface. If springs are encountered, drains must be provided to carry away the water. Foundation walls, being more or less porous, absorb and retain the surrounding moisture. They should extend, therefore, at least a foot above the surface so that this moisture may not be carried by capillary attraction into the wall above.

In soft ground piles must be driven. These are of wood where marine borers are not to be feared. They should be not less than 6 inches in diameter at the small end and straight. They will sustain 10 tons and if kept under water will not rot. The *Engineering News* formula for determining the safe load is $L = 2wh(Cs + 1)$, where L = safe load in pounds, w = weight of hammer in pounds, h = fall of hammer in feet, s = penetration of last blow in inches. Piles should never be driven with centers closer than 2 feet, 3 feet is much better. Concrete piles are either (1) made in place and driven, when cushioned head must be used; (2) filled in a hollow casing which is allowed to rust, or (3) filled in a casing which is withdrawn as the filling proceeds. They will carry 30 tons each. Concrete piles usually cost three to four times as much as wood.[1]

(31) Materials for Foundations.—For chemical buildings limestone or other rock easily attacked by acids should be avoided. The best material is usually cement concrete, made of broken stone or cinders, Portland cement and sand. The proportions used vary widely. Merritt & Co.'s handbook[2] recommends the following method for determining the proportions to be used when the materials have been chosen: "The proportion of matrix should slightly exceed the voids in the aggregate, so that each particle of the latter may be entirely covered with mortar. It is, therefore, evident that the proportion of voids in the aggregate to be used in each particular case should be ascertained in order to secure the best results. This is simply done by filling a vessel of known capacity, level full, with the loose aggregate, having previously thoroughly wet the latter. Then pour in as much water as the vessel will contain, and divide the volume of the water poured in by the volume of the vessel, the quotient will represent the proportion of voids. The mixture may then be proportioned accordingly, which will not only produce a better grade concrete, but frequently effect a material saving over the common method of arbitrarily adopting one of the usual proportions,

[1] Patton's Practical Treatise on Foundations; Patton's Treatise on Civil Engineering; Cortell's Allowable Pressure on Deep Foundations; The U. S. Public Works; Reinforced Concrete by Chas. F. March; Dock Engineering by B. Cunningham.

[2] Page 63.

regardless of the nature of the ingredients. Proportions in common use are by volume, 1 part cement, 2 parts sand and 5 parts stone, crushed to pass a 2-inch hole and free from fine dust, and 1 part cement, 2½ sand and 6 crushed stone. Where broken stone is not used, the following will give an idea of the proportionate amounts to be taken. These are the proportions used by the New York Rapid Transit Railway:

For brick masonry Column footing stones and rubble masonry..	1 portion Portland cement and 2 portions sand
Stone masonry	1 portion Portland cement and 2½ portions sand
Pointing	1 portion Portland cement and 1 portion sand

(32) Crushing Strength of Materials Used in Walls.—Walls are usually far stronger than is necessary. Where very heavy weights are to be carried, however, it is necessary that the engineer assure himself by calculation that they are sufficient. The following figures showing the crushing strength of materials commonly used will then be of assistance. Crushing strength of:

Brick, in lime and sand...........	8 tons per square foot
Brick, in cement....................	15 tons per square foot
Granite	75 tons per square foot
Sandstone	40 tons per square foot
Limestone	50 tons per square foot
Concrete	20 tons per square foot

(33) The Weight of Masonry in Pounds per Cu. Ft. is about as follows:[1]

Brick in mortar	100
Brick in cement.......................	112
Concrete	120–155
Masonry, dry rubble	140–160
Masonry, dressed	140–180
Mortar	90–100

(34) Brick Walls.—For chemical work these should almost always be laid in cement mortar, since lime mortar is rapidly eaten out by acids and acid gases. The brick should not be softer than the grade known as hard-stretcher and for damp situations very hard bricks are best. Most walls should be laid solidly, without

[1] Kent's Pocket Book p. 174.

air space and conforming to ordinary dimensions will be either 8½, 13 or 17½ inches thick. Since workmen *will* add lime to cement if allowed, they must be carefully watched to prevent this and several other evil habits, which if allowed impair the strength:

(1) It is a common practice to lay the brick dry or if they are wetted, to do it imperfectly. Bricks laid wet make a much better wall. If the men complain of sore hands, gloves should be provided.

(2) It is a common practice in laying brick to spread the mortar and daub a small bit on the edge as shown in Fig. 16. This leaves a void in the wall. The proper method is shown in Fig. 17, the brick being laid flat in the mortar and pushed against the preceding brick. In this way the mortar fills the lower part of the opening, the filling being completed when the next mortar course from above is laid. This method is more expensive, because slower, but gives a better wall.

(35) Laying Out Buildings and Building Walls.—At each corner of a building to be built, stakes are strongly driven in and connected by a cross-piece on which grooves marking the inside and outside of the walls are cut (Fig. 18). Care must be taken that these are not disturbed until the walls have reached such a height that they can be dispensed with. The engineer in charge should insist that corners be kept perfectly plumb by application of a plumb board shown in Fig. 19. To insure accuracy, lines should be stretched on the inside and outside of the wall and the thickness be tested as often as the line is moved. Bricklayers unused to these rigid requirements are apt to grow restive and considerable tact is needed to secure compliance.

Bricks are usually laid in what is known as American bond, as shown in Fig. 20, in which every eighth course consists of headers, the intervening courses being stretcher courses as shown. In English bond there are alternate courses of headers and stretchers. Flemish bond has alternate headers and stretchers in each course, as shown in Fig. 21, and this is quite frequently used in office buildings. Here the headers may be slightly different in color to give variety. The pointing mortar used is sometimes colored to match the brick by adding iron ores; black, by adding charcoal or white by mixing lime and marble dust. Nowadays, stone masonry is little used except in foundations and retaining

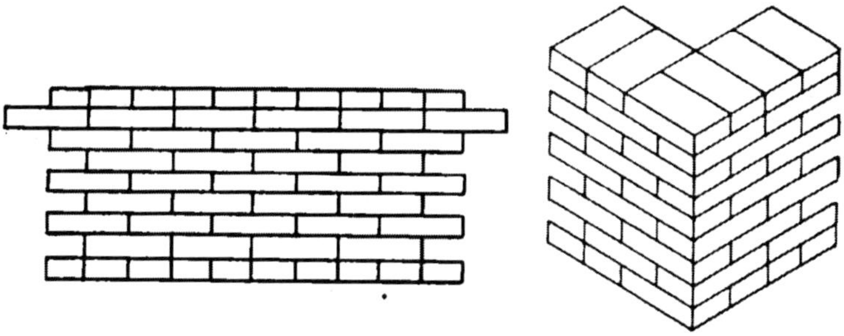

Fig. 20.—American Bond.

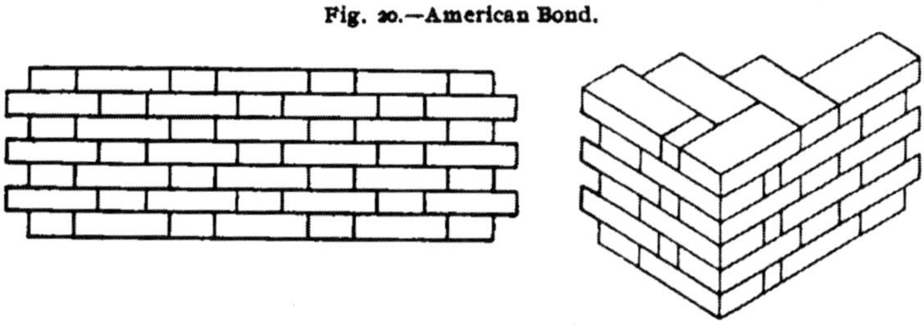

Fig. 21.—Flemish Bond.

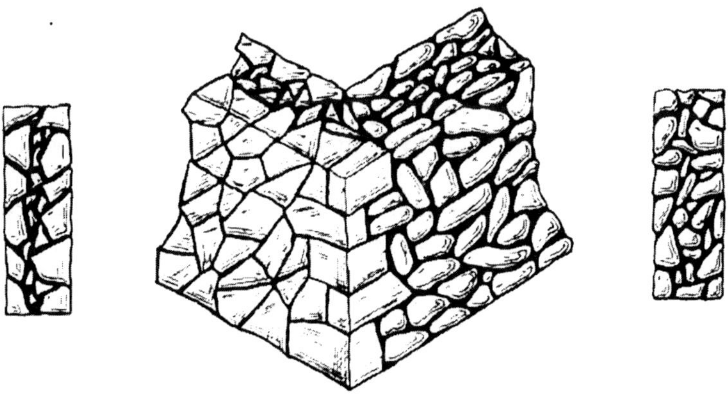

Irish Wall. —Fig. 22.— Rubble Wall

walls. It may sometimes be advantageously used where it is desired to dispose of excavated material. Stone with flat surfaces may be laid up rapidly by good masons. It is desirable, where much of this work is to be done, to keep a careful watch on the work done by each man, both with respect to the quantity and quality of his work, and to promptly lay off those who do not come up to the mark in either respect. Work not properly and carefully bonded should generally be torn down. Cement should be mixed in batches of moderate size and be used promptly before initial set occurs. All work of this character must be continuously watched. Illustrations of stone work with proper bonds and corners are shown in Fig. 22. Walls are often laid with brick or stone corners which must then be properly bonded as shown in Fig. 23, which also shows string courses and a window opening. This is built of Ashlar or dressed stone.

(36) Hollow Tile may be used for walls and floors as shown in Fig. 13a.[1] Wooden floors are laid as shown in Fig. 13b.[2] The Johnson floor system shown in Fig. 13c has been used on spans up to 25 feet. This is believed to be the strongest and lightest fire proof floor made. Segmental arch floors between I beams with or without tie rods which may be bare or protected may also be used. These are especially useful for long spans. The method of construction with tie rod protected is shown in Fig. 13c. The segmental arch construction makes the cheapest floor.

These blocks are laid in one part cement and three parts sand. Lime, not to exceed 10 per cent. by volume may be allowed in the mortar. The stucco covering is for first coat: one cement, one-tenth lime, two sand; second coat: one cement, one-tenth lime, two and one-half sand; third coat: one cement, one-tenth lime, three sand. Before applying the first coat the surface must be wetted thoroughly and applied with force so as to form good keys. Last coat of slimes should be waterproofed.

(37) Safe Load for Hollow Tile.—A hollow tile wall or column thick enough to render danger of buckling negligible will sustain a load of 300 tons per square foot of area. With a factor of safety of five this will allow us to load to 60 tons per square foot.

[1] National Fire Proof Long Span Fire Proof Cons., p. 25, lower fig.

[2] Standard Fire Proof Cons. Nat. F. P. Co., p. 21, lower fig.

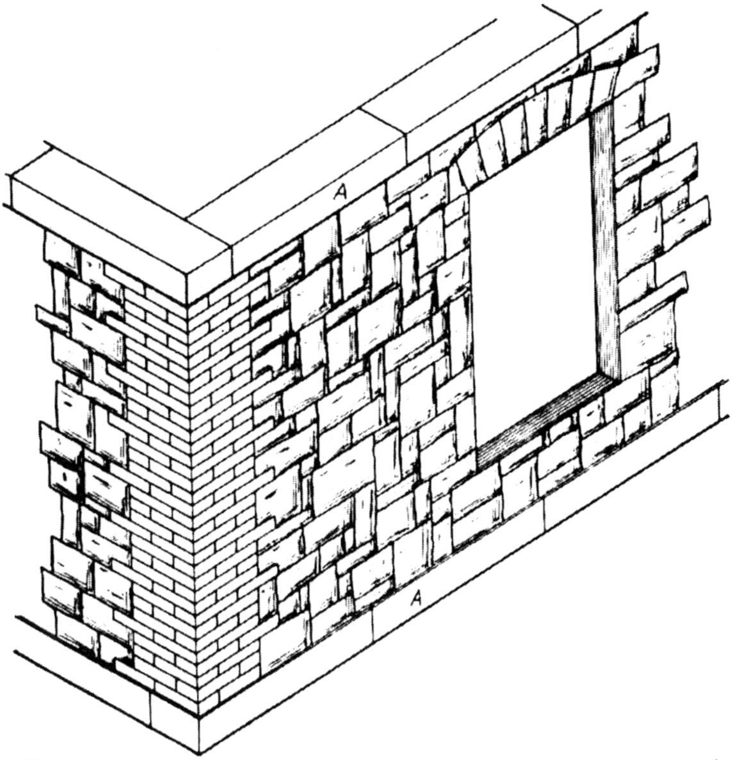

Fig. 23.—Ashlar, or dressed stone wall, with string courses, (A), window opening, and brick corner.

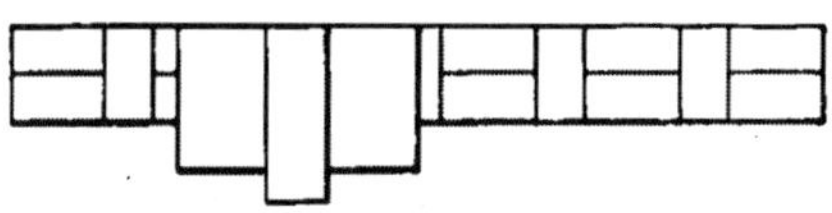

Fig. 24.—Plan of pilaster, with bearing-plate of cast-iron, and steel floor-beam.

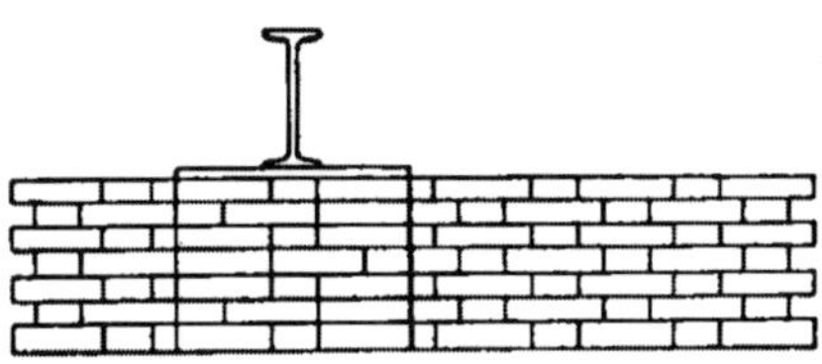

Fig. 24a.—Elevation of 24.

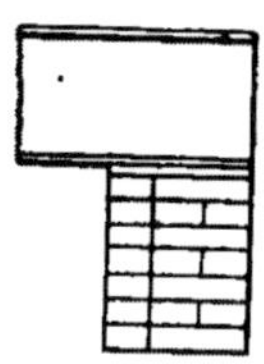

Fig. 24b.—Section of 24.

(38) **Pilasters** are often used to strengthen a wall and to increase the bearing area under beams. When pilasters are used, the number of corners is increased and thereby the cost of laying the walls. This additional cost sometimes more than balances the brick saved. A wall built with pilasters is shown in Fig. 24.

(39) **Floors** may be made of concrete, but should be supported by exposed steel beams, which must be frequently inspected and painted with asphaltum paint. Concrete makes an excellent floor except where exposed to acid laid upon the ground. The earth should first be excavated for a foot, and rammed, and a foundation of stone and cinders well rammed be first laid. Upon this, four inches of concrete and one inch of cement and sand (one cement, two sand) is spread. This must not be allowed to freeze.

For second story floors, steel or yellow pine beams well bridged may be used and as a covering, yellow pine flooring. The resin contained in this wood makes it resistant to acids and it has the advantage also of cheapness.

(40) **Roofs** should be laid nearly flat in most cases and covered with gravel asphalt by the Merritt system. Where composite (cement asbestos) tiles are used steel truss supports well painted may be used. Tin, slate and wooden roofs will not answer.

Fireproofing most chemical construction does not yet appear to be practicable because concrete is permeable to liquids and gases and the reinforcing steel rods might give way. This has not, however, had a sufficient test.

The tops of walls should be carried above the roofs and, when practicable, dividing walls should be without openings. The tops of walls should be covered with a coping made of overlapping tile laid in cement. Cement copings may also be used.

(41) Weight of Wood.

	Weight in lbs. per cubic foot of dry wood		Weight in lbs. per cubic foot of dry wood
Ash	39–45	Larch	35
Beech	45–46	Lignum vitæ	62
Birch	41–47	Linden	37
Cedar	31–39	Locust	42–46
Cherry	41	Mahogany	51
Chestnut	35	Maple	38–43
Cypress	29–33	Oak	45–59
Dogwood	47	Osage orange	48
Ebony	76	Pine, white	24–28
Elm	34–45	Pine, yellow	38
Fir	32–37	Poplar	29–30
Gum	37–57	Spruce	21–28
Hackberry	45	Sycamore	35–37
Hackmatack	37	Tamarack	46
Hemlock	24–26	Teak	51
Hickory	47–56	Walnut	36–38
Holly	47	Willow	34
Hornbeam	47	Redwood	26
Juniper	35		

(42) Beams and Girders.[1]—Safe load on Beams. Uniformly distributed load,

$$\text{Safe load in lbs.} = \frac{2 \times \text{breadth} \times \text{square of depth} \times A}{\text{span in feet}}$$

$$\text{breadth in inches} = \frac{\text{span in feet} \times \text{load}}{2 \times \text{square of depth} \times A}$$

The depth is taken in inches. The coefficient A is one-sixteenth the maximum fiber stress for safe loads, and is the safe load for a beam one inch square one foot span. The following values of A are given by Kidder as one-third of the breaking weight of timber of the quality used in first class buildings. The values for stones are based on a factor of safety of six.

[1] Kent's Mechanical Engineers Pocket Book, p. 1335.

VALUES FOR A COEFFICIENT FOR BEAMS.

Cast iron	308
Wrought iron	666
Steel	888

American woods—

Chestnut	60	Spruce	70
Hemlock	55	Poplar	65
White oak	75	Hudson River blue-stone flagging	25
Georgia yellow pine	100	Granite (average)	17
Oregon pine	90	Limestone	14
Red or Norway pine	70	Marble	17
Eastern white pine	60	Sandstone	8–11
Western white pine	65	Slate	50
California redwood	60		
Texas yellow pine	90		

CHAPTER III.

BOILERS.[1]

A great variety is to be found in boilers in actual use depending upon the water and the fuel available. Where fuel is cheap, boilers with few tubes are used and this is also necessary in some cases where the water scales badly. Three types will be described here:

(43) Internally Fired Upright Tubular Boilers.—These are generally used for small power and are constructed as shown in Fig. 25. The upper ends of the tubes, which are usually about two inches in diameter, project into the steam space and as a result the steam is slightly superheated. The tubes of this and other boilers are expanded into the crown and top sheets by means of an expanding tool, as shown in Fig. 25a. Such boilers are much used by contractors and are often mounted on wheels with engine and pump attached. Upright boilers may be had from 24-54 inches diameter, 4-10 feet high and with a rated horse-power of 4-60.

(44) Horizontal Tubular Boilers.—Where fuel is cheap, single tube boilers like the Cornish (Fig. 27), or Galloway boilers (Fig. 28) may be used. With higher priced fuel for moderate sized installations some form of locomotive boiler is preferred.

A many tube boiler with furnace is shown in Fig. 26 and 26a. They have 3- or 4-inch tubes, a diameter of 37-72 inches and a length of 8-20 feet, and a horse-power capacity of 15-200. In this boiler the products of combustion pass under the boiler from left to right, then through the tubes right to left, and then back over the boiler to the chimney. The man-holes shown are for cleaning and are 11x15 inches. Detail shown in Figs. 28 and 29. The best grate is the herring bone shown in Fig. 30, and as applied to an upright boiler in Fig. 31. Figs. 32 and 33 show boiler settings and a furnace door. These doors are sometimes made with water jackets.

[1] Barr, Boilers and Furnaces; Hutton, Mechanical Engineering of Steam Power Plants Steam, Its Generation and Use, and The Stirling Water Tube Boiler, published by the Babcock and Wilcox Co.

UPRIGHT BOILER.

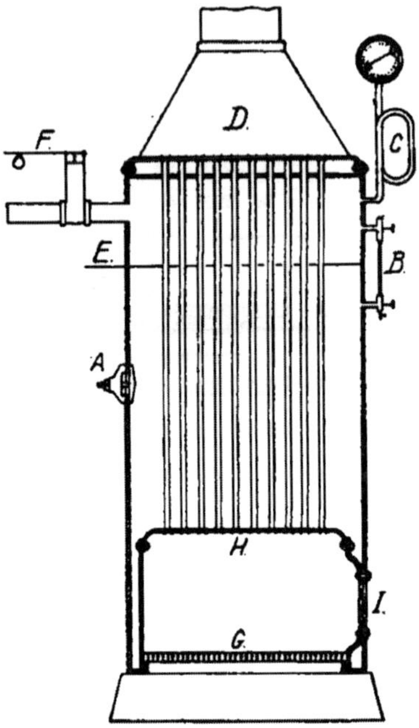

Fig. 25.

A. Hand-hole for cleaning.
B. Water Gage.
C. Pressure Gage.
D. Hood.
E. Water line.
F. Safety-valve.
G. Grate.
H. Crown-sheet.
I. Fire-door.

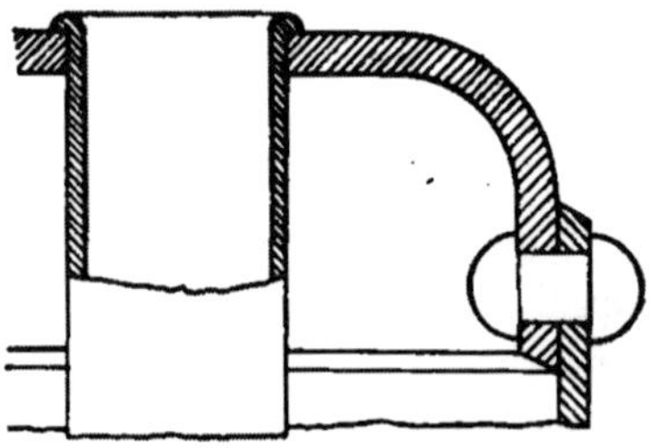

Fig. 25a.—Boiler tube expanded into top sheet and upset.

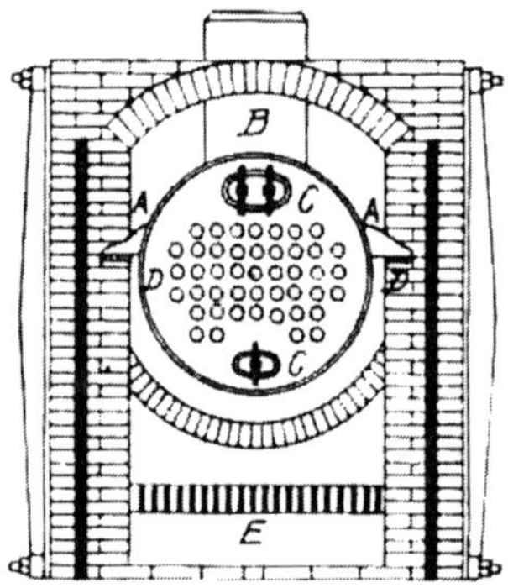

Fig. 26.—Horizontal Boiler.

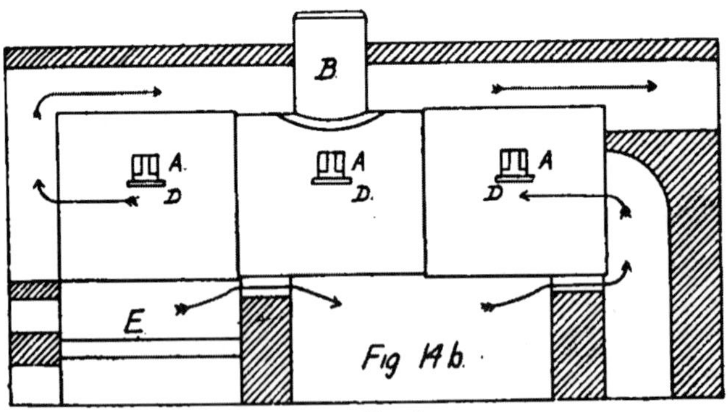

Fig. 26a.

A.A. Cast-iron lugs to support boiler.
B. Steam dome.
C.C. Man-holes.
D. Cast-iron plate.
E. Grate.

THE CORNISH BOILER.

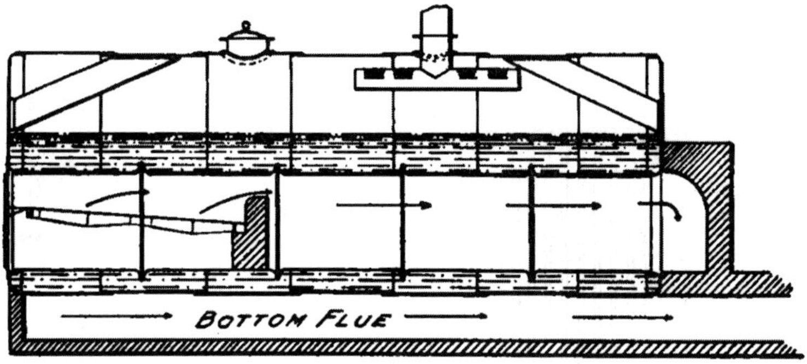

Longitudinal Section A-A.

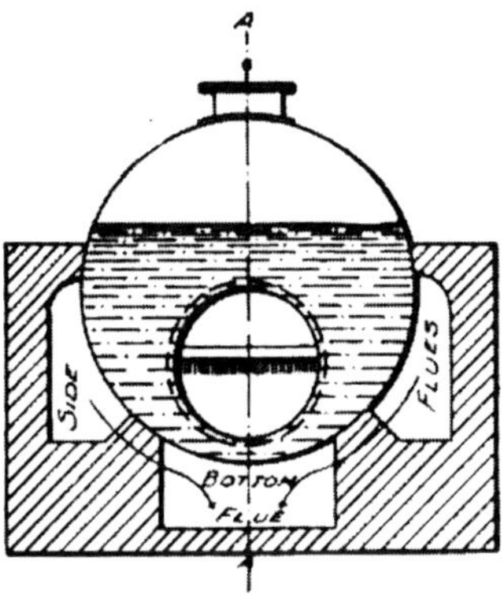

END VIEW

Principal Dimensions:

Diameter of shell	3'-6" to 6'-0"
Length	8'-0" to 22'-0"
Diameter of flue	2'-2" to 3'-6"

The Cornish Boiler consists of an outer shell within which is a flue of sufficient size to permit its being fitted with a furnace for the combustion of the fuel. The gases from the combustion chamber leave the flue at the rear, divide and return along the sides of the boiler to the front where they reunite and pass through a flue under the boiler to the rear of the shell and into the chimney.

THE ADAMSON JOINT.

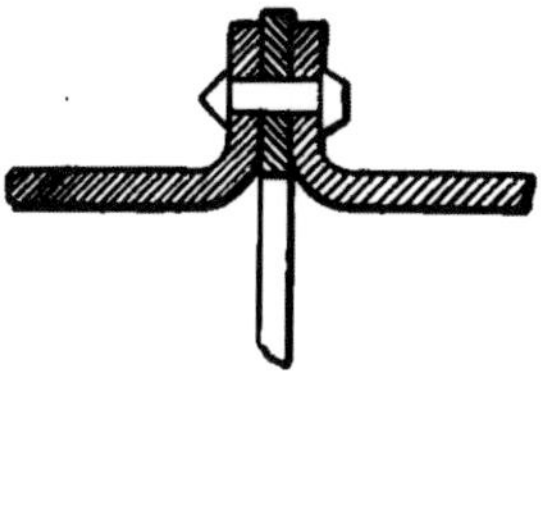

THE ADAMSON FLANGED SEAM.

This seam insures sufficient longitudinal flexibility to prevent destructive strains within the boiler and the flanged ring makes it more able to resist collapse.

Fig. 27.

The Galloway and Lancashire Boilers.

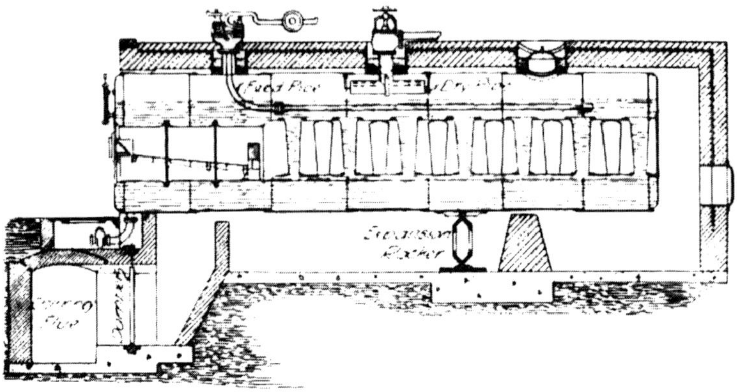

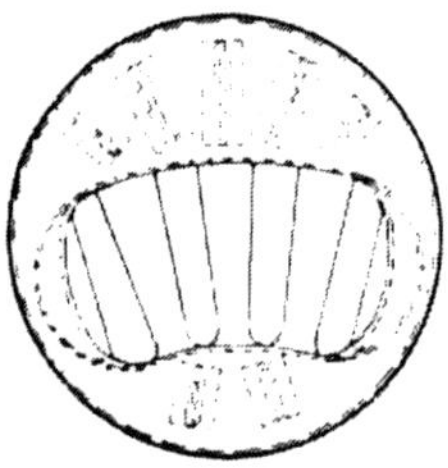

Galloway Boiler.
End View.

The Galloway Boiler.
Longitudinal Section.

The Galloway Boiler is fitted with two furnaces merging into a segmental combustion chamber. This chamber is fitted with tapered water-tubes which increase the effective heating surface, promote a better circulation of the water, and act as stays, thus greatly adding to the strength of the flue.

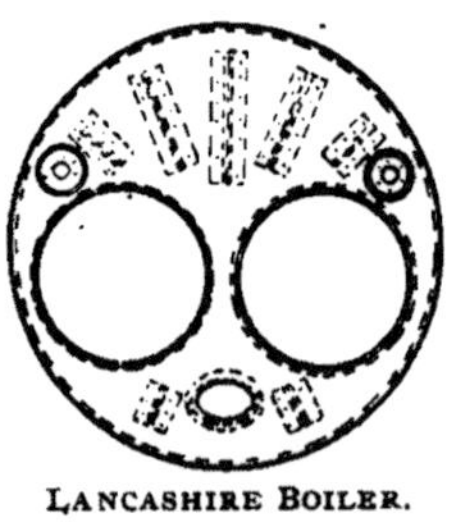

Lancashire Boiler.
End View.

The Lancashire Boiler.

This boiler is similar to the Cornish type, except that it is fitted with two furnaces and two flues. Thus lateral rigidity and stiffness are secured.

Fig. 28.

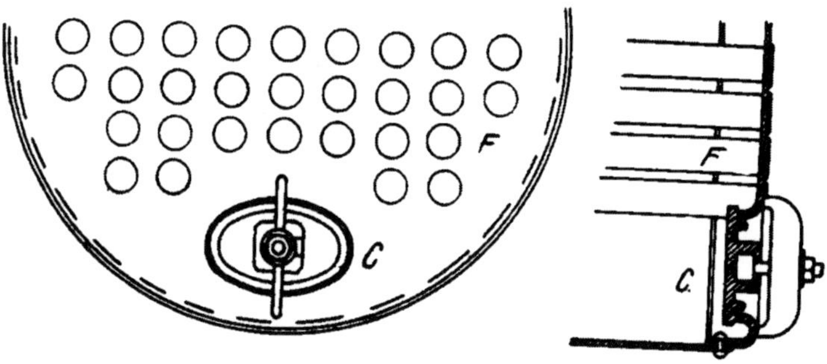

Fig. 28a.—Detail of Fig. 26.
C. Man-hole. F. Boiler tubes.

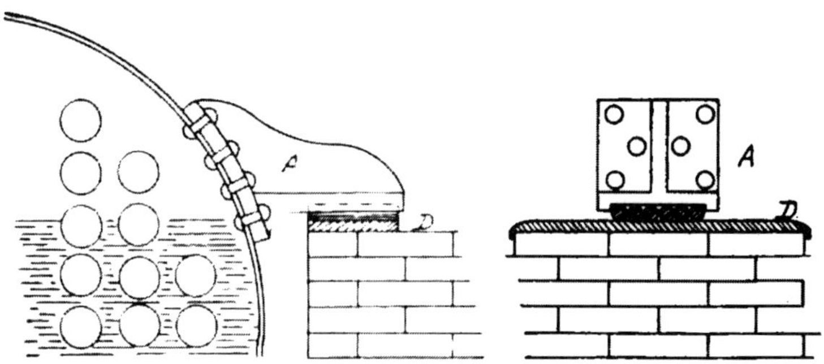

Fig. 29.—Boiler Setting.
A. Cast-iron lugs supported by rollers on cast-iron plate D.

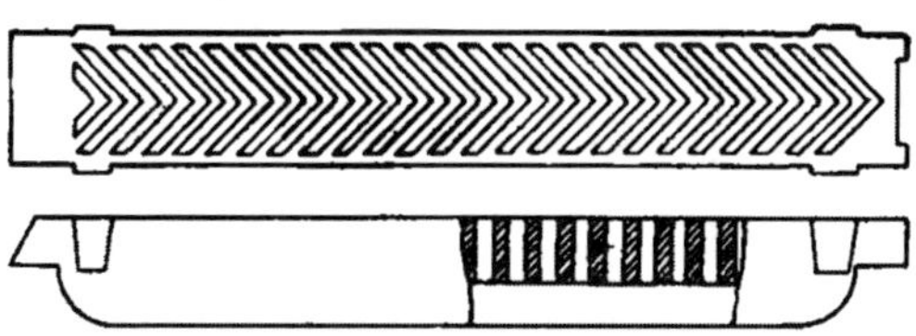

Fig. 30.—Herring bone grate.

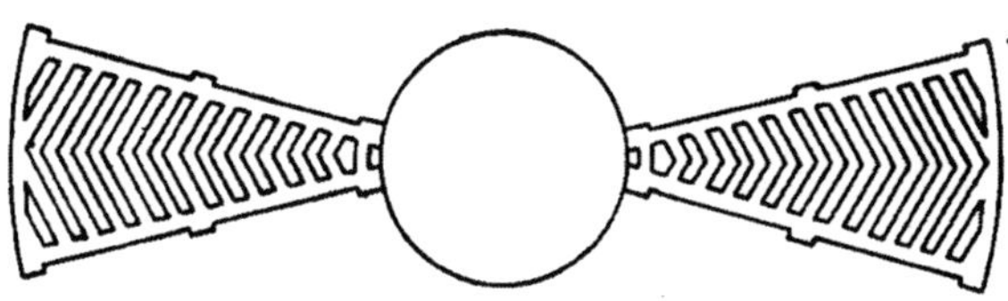

Fig. 31.—Herring bone grate for upright boiler.

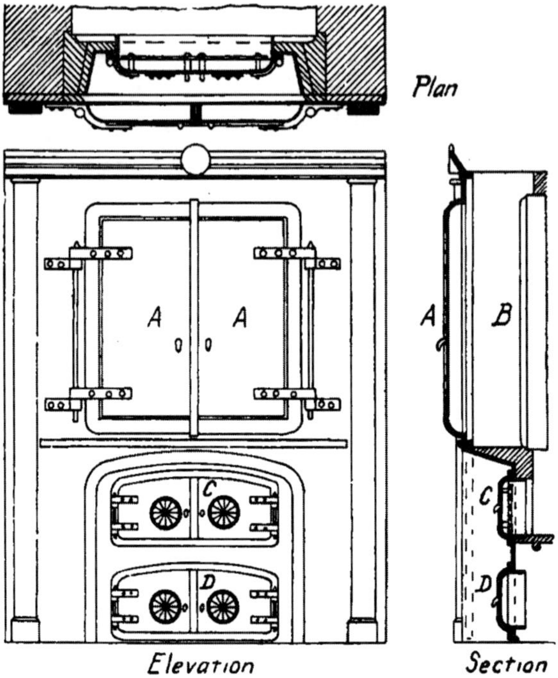

Fig. 32.—Boiler setting.
AA. Boiler doors. C. Fire door.
B. End of boiler. D. Ash door.

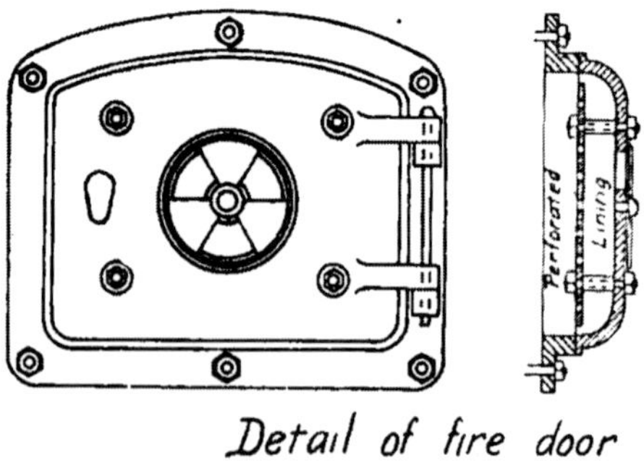

Fig. 33.—Furnace door.

(45) Water Tube Boilers of which the Babcock and Wilcox is shown in Fig. 34. This consists of a series of 4-inch seamless steel tubes expanded into serpentine shaped cast iron or wrought steel headers according to the pressure to be carried. A header is shown in Fig. 35 and in detail in Fig. 36. Proper circulation in such boilers is essential both for safety and economy. These boilers have one or more longitudinal or cross drums above the tubes and a small mud drum below and in the rear. They are specially fitted for high pressures.

Boilers are tested under hydraulic pressure and the strain put upon them should never be allowed to exceed more than half the tested limit. They should be inspected after running 1000 hours and tested whenever this seems necessary, especially after having been used for some time or after extensive repairs. A good boiler will evaporate 10 pounds water for every pound of fuel burned and often exceed this, depending upon the quality of the fuel.

Fig. 37 shows the water gage and gage cocks. Fig. 38 shows detail of the glass gage and Fig. 39 of the gage cocks.

Boilers are also provided with a steam gage, shown in Fig. 40, and a safety gage in which the safety valve is held down by a weighted lever as shown in Fig. 41, or by a spring, as shown in Fig. 42. Check valves (Fig. 43) are inserted in the feed water pipe to prevent the water being forced back by the steam pressure, and blow-off valves (Fig. 44) are provided to allow the removal of mud and scale. Some of the necessary qualifications of a boiler are as follows:

1. The boiler must have sufficient capacity for the work to be done so that neither the supply of steam nor its pressure shall fluctuate seriously. At the same time the water may be kept at constant level.

2. Enough surface to the water so that foaming need not occur.

3. Thorough circulation. Water is a poor conductor of heat and good circulation spells economy in coal, and prevents local over-heating which injures the metal; it also reduces boiler incrustations.

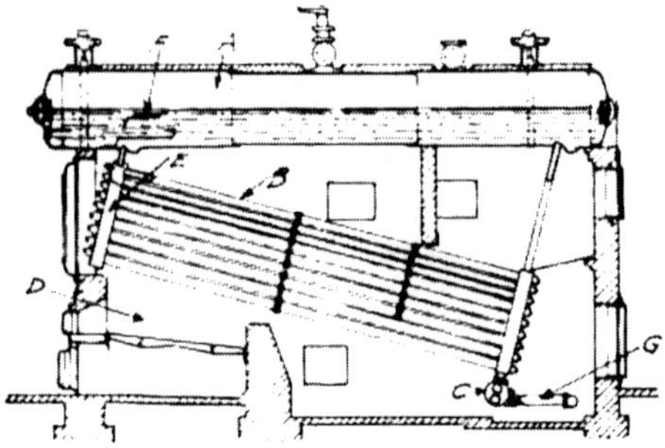

Fig. 34.—Water Tube Boiler.
A. Drum. B. Tubes. C. Mud drum. D. Combustion Chamber.
E. Headers. F. Deflector. G. Blow Off.

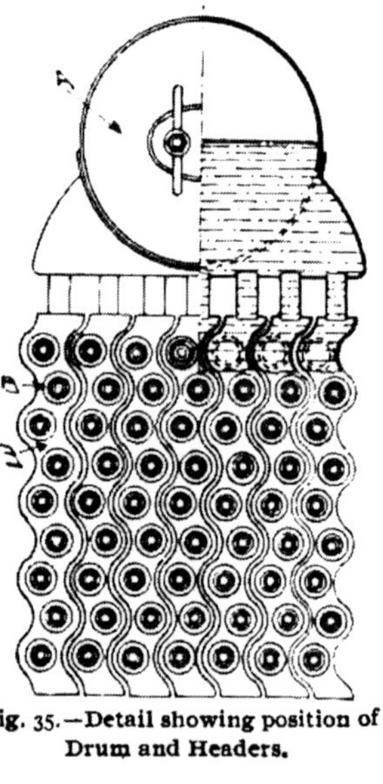

Fig. 35.—Detail showing position of Drum and Headers.

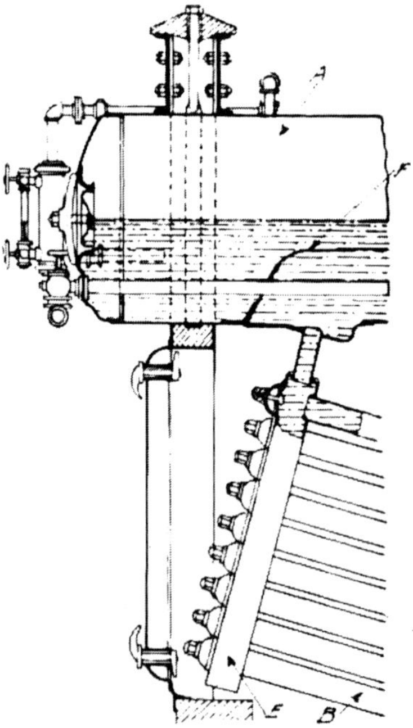

Fig. 35a.—Detail showing Front of Drum and Headers.

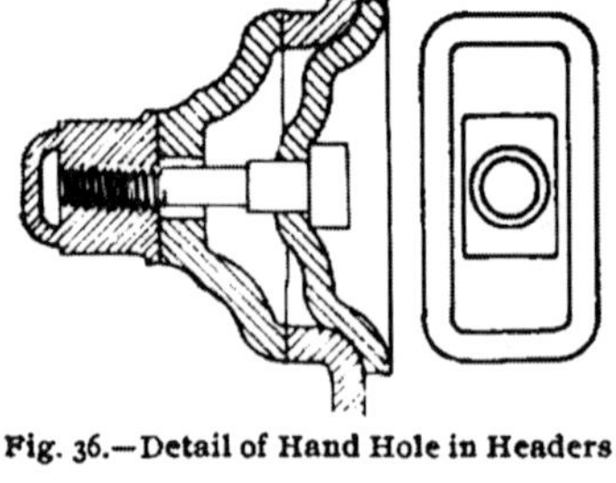

Fig. 36.—Detail of Hand Hole in Headers.

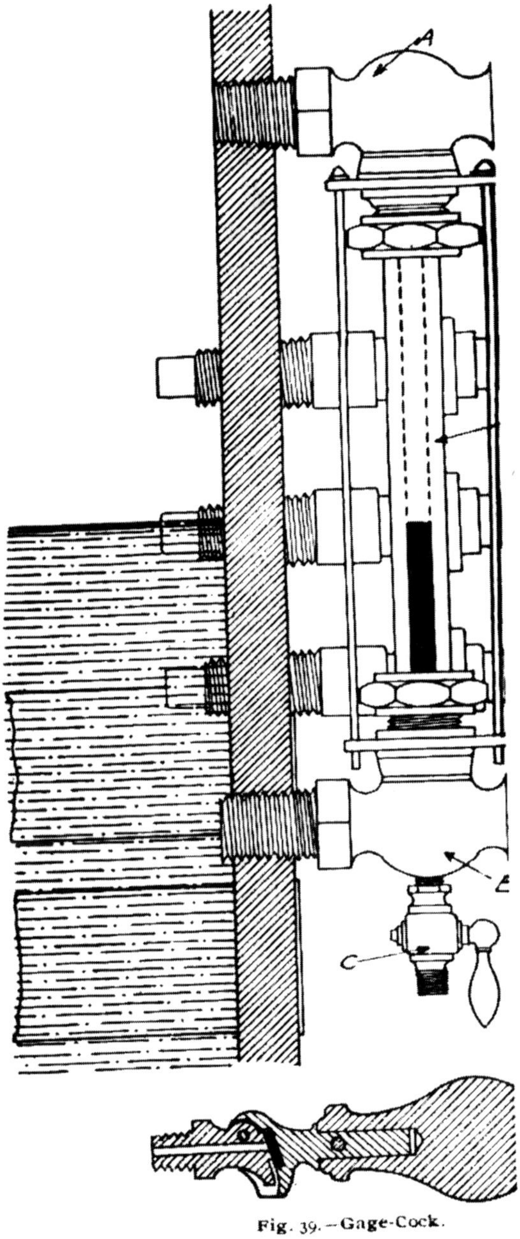

Fig. 39.—Gage-Cock.

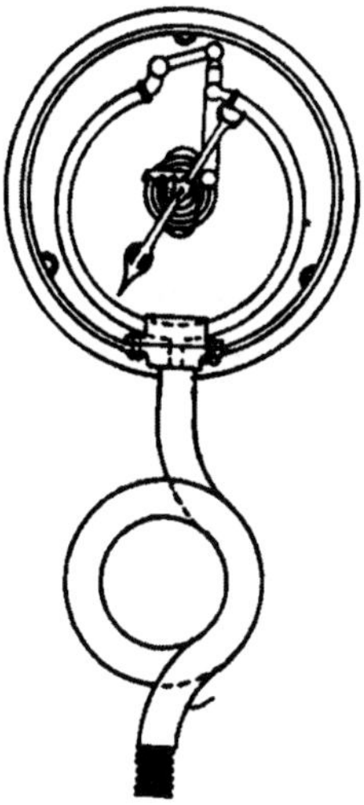

Fig. 40.
Steam Gage with Siphon.

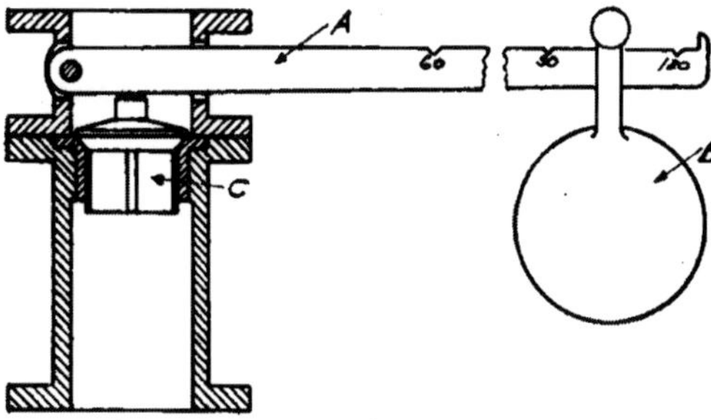

Fig. 41.—Combined Safety and Stop Valve.
A. Lever. B. Weight.
C. Valve.

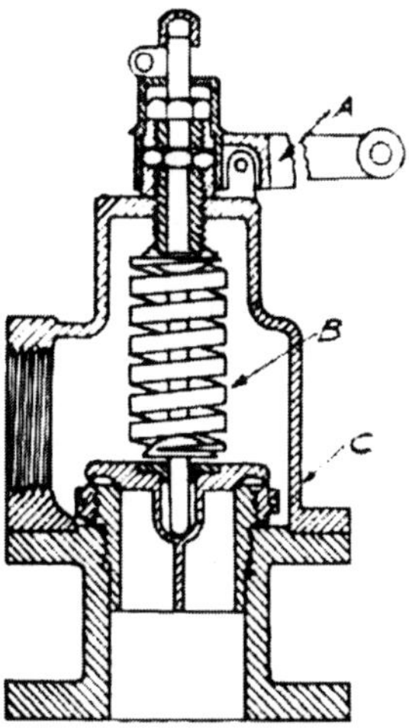

Fig. 42.—Safety Valve.
A. Lever. B. Spring. C. Case

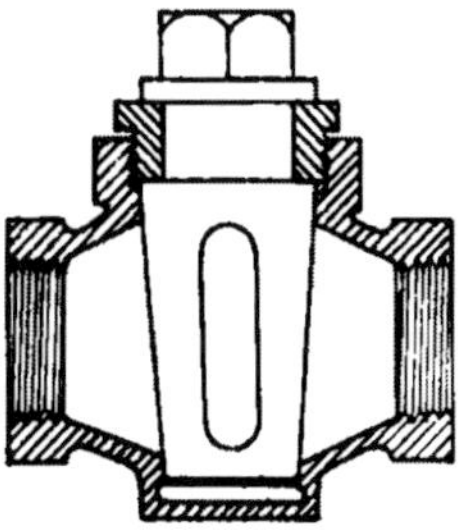

Fig. 44.—Plug-cock.

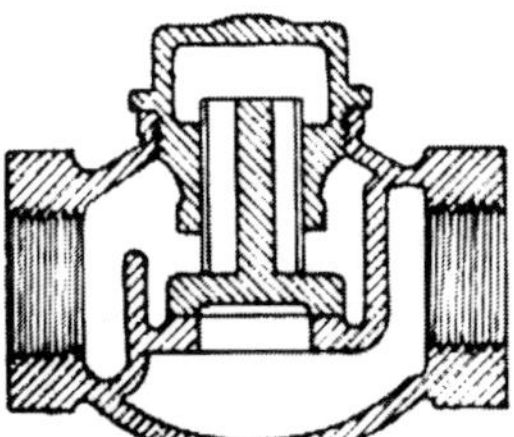

Fig. 43.—Check-Valve.

4. A combustion chamber where the gases attain high temperature ensuring freedom from smoke and maximum economy. A brick lined combustion chamber with step or Siemens grate or automatic stokers is best for large boilers burning bituminous coal.

Tube boilers are best for economy, since they present a large heating surface and are less liable to cause serious explosions, but they cost more. The chief cause of scale in boilers is either calcium carbonate or sulphate or both. The substances mostly used in purifying and softening water for boiler purposes are lime, soda ash and tribasic sodium phosphate. Concerns are now in existence which analyze boiler water and furnish mixtures of chemicals guaranteed to stop scaling. Magnesium chloride is a very troublesome constituent of boiler water, as it is apt to cause pitting. As the dissolved substance in water remains in the boiler, such residues soon contain a concentrated solution of dissolved salts; boilers in constant operation should, therefore, be blown off at least once a week.

The feed water may, in small boilers, be supplied by means of injectors; but these are apt to be troublesome, since small pieces of refuse will clog them. It is more expensive, but much more satisfactory and in the long run probably cheaper, to install a direct acting steam pump. This should not be one of the smallest sizes made, which not only look but often act like a toy. For gritty feed water plunger (Fig. 45), and for ordinary clear water the Worthington piston pump, shown in Fig. 46, is well adapted.

It is worth considering whether it may not, in some instances, be desirable to use forced draft fans actuated by electric motors for supplying air to boilers? Such positive regulation, while adding to the cost of operation, is economical of coal since the draft is under complete control and the amount of air passing through the fuel may be so regulated as to supply just the amount of oxygen needed for complete combustion. Operating in this way the cost of tall chimneys may often be avoided. A carbon dioxide regulator installed in the flue and a bonus paid to firemen often bring good results.

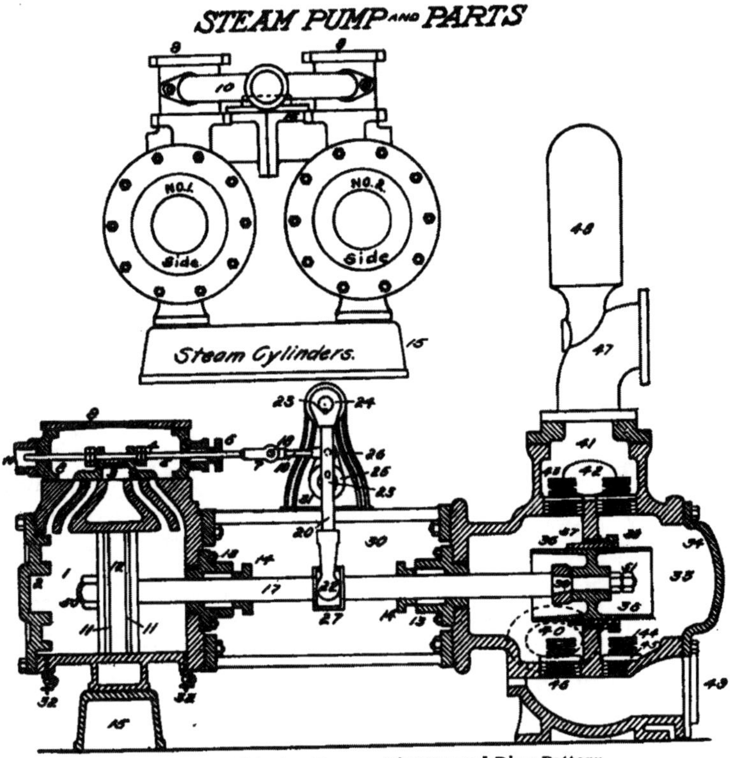

Fig. 45.—Worthington Pump. Plunger and Ring Pattern.

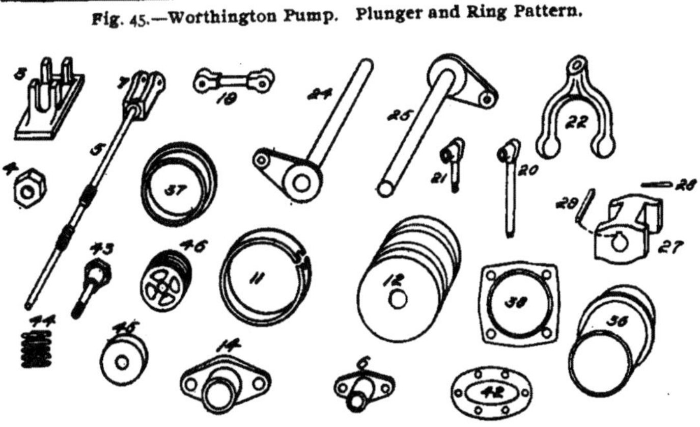

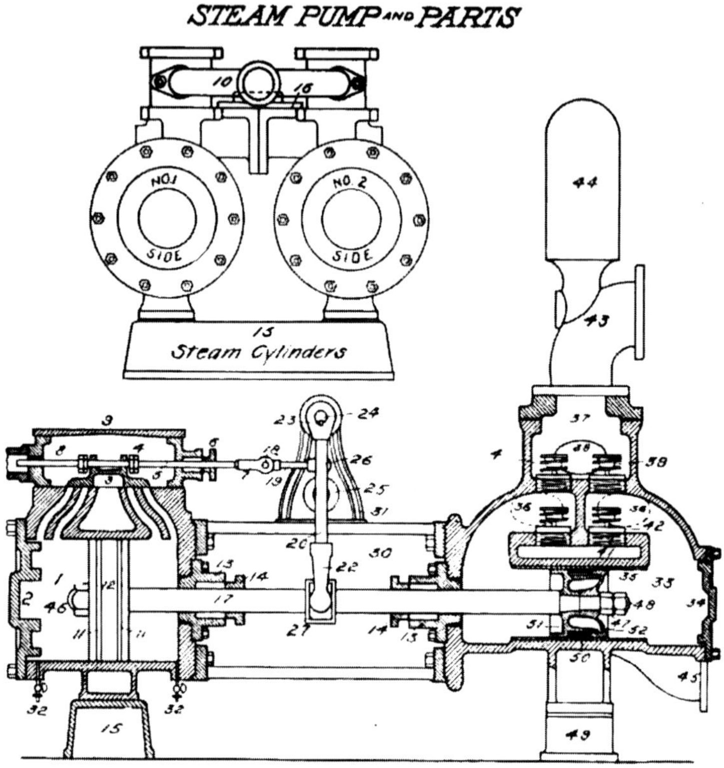

Fig. 46.—Worthington Pump. Piston Pattern.

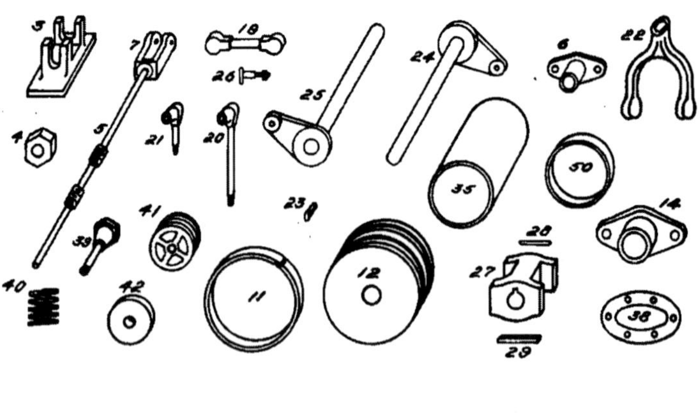

(45a) Surface Combustion Boilers.—Bone has shown[1] that gas and air mixed in proper proportions and burned in contact with refractory materials give an almost flameless combustion and produce a large amount of radiant heat. This radiant heat is rapidly and almost completely absorbed by metal surfaces, such as occur in ordinary boilers. Boilers constructed on this principle have been introduced which give very high efficiency and very rapid rates of evaporation. The most recent discussion of this boiler is contained in "Coal and Its Scientific Uses," London, 1919, to which the reader is referred.

(45b) Heat Transmission in Boilers.—The heat transmitted through the tube of a boiler is of three sorts: (1) radiant heat, (2) conducted heat and (3) heat of convection. The radiant heat is greatest in amount when flameless combustion is produced [see (45a)]. The amount of heat conducted is greatly decreased by the formation of a "dead film" of gas on the one side and a similar film of water on the other side of the boiler tube. These films greatly decrease the conductivity and are the cause of heavy loss; their effect is best decreased by a high velocity of the gas on one side and of water on the other. This rapid motion also greatly increases the absorption of heat due to convection.

[1] Bone, *Proc. Amer. Gas Inst.*, VI, 565; *J. Frank. Inst.*, 173, 101; *Ber. der Deutsch. Chem. Ges.*, 46, 5.

CHAPTER IV.

PRIME MOVERS.

The source of power may be either a water wheel or turbine, steam or gas engine or electric motor. The latter is indeed not a prime mover since the current is usually produced by one or the other of the two first named.

(46) The Steam Engine.—When steam under pressure is admitted behind a piston (Fig. 47), the piston is driven before it until the other end of the cylinder is reached. By means of the piston rod working through the stuffing box, C, and attached to the cross-head, D, and connecting rod, E, this motion is communicated to the fly wheel, F. If at the end of the stroke, the supply of steam is cut off and another port opened into the air, the momentum of the wheel, F, will carry the piston back to its original position when steam may be again admitted.

(1) The above described engine is a SINGLE-ACTING, NON-CONDENSING ENGINE. If the steam escaping from the cylinder is passed into a vacuum chamber cooled by a water spray, we have a

(2) SINGLE ACTING CONDENSING ENGINE. If the steam is applied alternately to either end of the piston, we have, if the exhaust steam escapes into the air, a

(3) DOUBLE ACTING NON-CONDENSING ENGINE, and if into a cooled vacuum, a

(4) DOUBLE ACTING CONDENSING ENGINE. In this engine the working parts are actuated by an eccentric on the main shaft, and serve to allow the steam to enter and escape at the proper moment automatically. This is illustrated in Fig. 48.

(47) Expansive Working.—When high pressure steam is used it is not necessary to allow steam to enter during the entire stroke. By adjusting the slide valve, the supply may be cut off during any part of the stroke, and the expansive action of the steam will carry the piston to the end of the cylinder. In this case, however, the pressure in the cylinder must decrease after the steam is cut off. The horse-power of a given engine is always less when it works expansively and the economy of fuel greater.

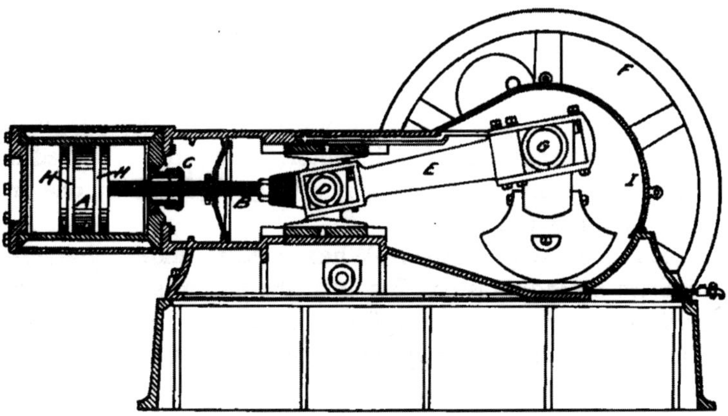

Fig. 47.—Steam Engine.

A. Piston.	F. Fly Wheel.
B. Piston Rod.	G. Eccentric.
C. Stuffing Box.	H. Packing Rings.
D. Crosshead.	I. Oil Guard.

E. Connecting Rod.

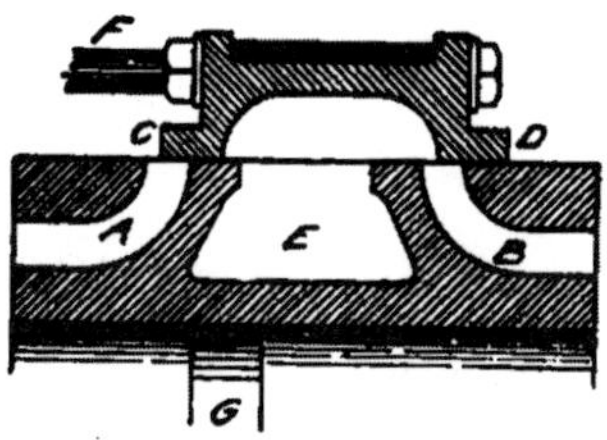

Fig. 48.—Slide Valve.

A. Left Port.
B. Right Port.
C-D. Lugs on Valve.
E. Exhaust Port.
F. Rod.
G. Piston.

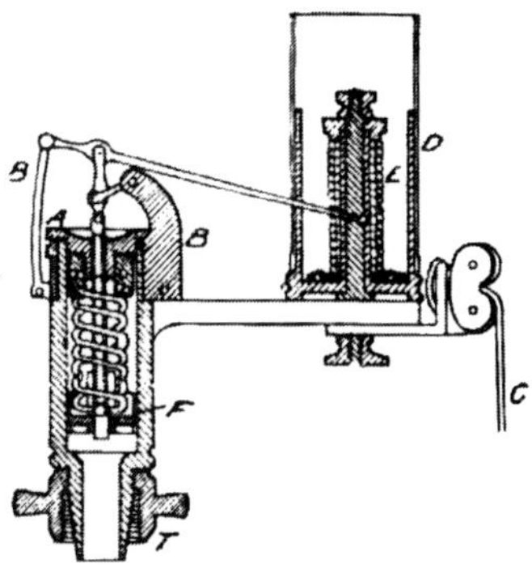

Fig. 49.—Crosby Indicator.

A. Cap.	D. Drum.
B. Linkage.	E. Spiral Spring.
C. Cord.	F. Piston.

T. Thread.

(48) Horse Power.—This is a unit devised by James Watt, the inventor of the steam engine, to measure its performance in terms comparable to that of the source of power then most commonly used. As Boulton & Watt manufactured engines, they desired that the unit should be fully equal to the performance of a powerful horse. It was found that a heavy draft horse could do work corresponding to the lifting of 33,000 pounds one foot in one minute, and this was the unit chosen. The French unit "force de cheval" = 0.95363 English horse-power or one English horse-power = 1.0163 "force de cheval."

Horse power[1]	Eng. ft. lbs. per minute	French kil. meter per minute	Austrian ft. lbs. per minute
English and American...	33,000	4,572.9	25,774
French	32,4704	4,500	24,561
Austrian	33,1342	4,5495	25,800

One kilowatt hour = 1.34 H. P. One English H. P. = 746 watts.

(49) The Work Done By an Engine[2] is equal to the area of the piston in square inches multiplied by the mean effective pressure per square inch multiplied by the length of the stroke in feet multiplied by the number of strokes per minute.

(50) Heat and Work.—Joule, and later Rowland, have determined that the amount of heat produced by 778 foot pounds is equal to the amount of heat required to raise one pound of water at 39° F. to 40° F. or the British thermal unit. This theoretical relation is never attained in practice, but serves as the highest attainable amount by which to measure the operations of a power plant, the boiler and engine.

(51) Indicator Diagrams.—If we call the pressure on the piston P, the length of stroke L, the area of the piston A, the number of strokes per minute N, then PLAN will be the work performed and $\frac{PLAN}{33,000} = H.P.$ In any one engine working normally L and A are fixed and, if the pressure in the boiler does not vary rapidly, N will be nearly a constant. During each stroke, however, P may change rapidly and a very clear idea of the en-

[1] From Hutton.

[2] Ganot's Physics, Section 482.

gine's performance may therefore be obtained by a study of indicator cards.

(52) The Indicator Piston invented by James Watt and shown in Fig. 49, is put into free communication with the engine cylinder by means of the thread T, so that variations in the pressure are recorded on the cylinder C covered with paper. The area of the diagram obtained in this case represents the actual work done at each stroke and the form, to the practical eye, is a sufficient measure of the efficiency of the engine. Fig. 50 shows such an indicator diagram. This may be divided into an equal number of parts by drawing perpendiculars through it. The height of each perpendicular is measured; the total divided by the number gives the average pressure which, multiplied by the area of the piston, length of stroke and number of strokes per minute gives the work done.

(53) Economy of the Engine.—The British Thermal Units evolved in the burning of a pound of average coal are, say, 12,000. The theoretical equivalent of 1 horse-power hour in British thermal units equals $\frac{33,000 \times 60}{778} = 2545$. One pound of coal should, therefore, produce about 5 horse-power. The usual consumption of coal is from 1.25 to 3 pounds per horse-power hour so that only about one-sixth to one-fourteenth of the theory, say 17 to 7 per cent., is attained in practice.[1]

(54) The Piston.—A modern form of piston is shown in Fig. 51, which shows the connection with the piston rod and the piston rings. These are made of steel and are divided by the cut shown in Fig. 52 or 53, that they may spring against the cylinder when in use so as to make a steam tight joint. This is further ensured by so placing the joints on the piston head that the small leak through the first ring is opposite a solid portion of the second ring.

(55) The Stuffing Box.—This is made as shown in Fig. 54, and properly packed. Its operation is understood by inspection of the drawing.

[1] See also an article by Alex. Dow, *Trans. Amer. Electrochem. Soc.*, **32**, 65, 1917.

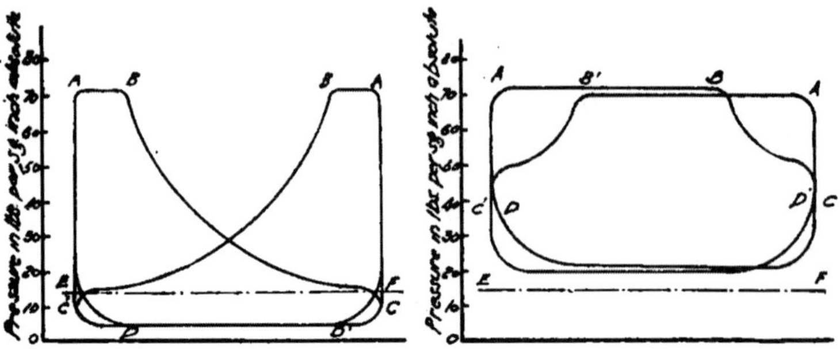

Fig. 50.

a. The steam is expanded about seven times and as the engine is a condensing one the back pressure is about 5 pounds, the pressure during admission being about 75 pounds.

b. A non-condensing engine, the back pressure being slightly above atmospheric, here the steam is cut off at about ¾ of the stroke.

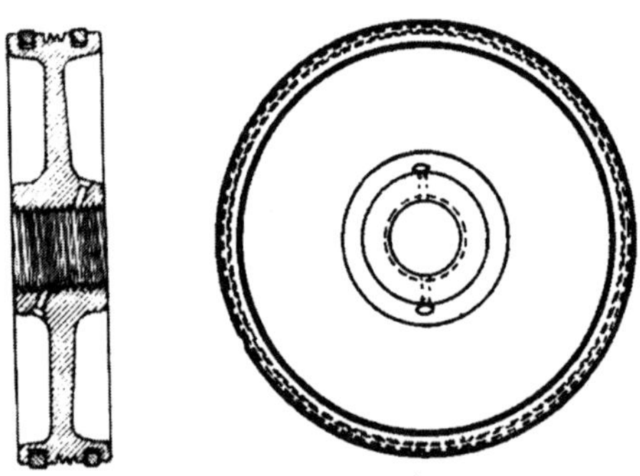

Fig. 51.—Steel-plate Piston.

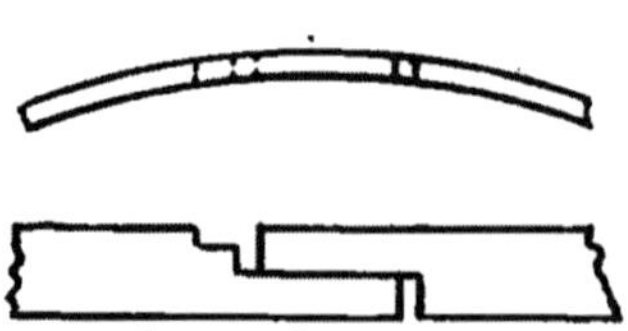

Fig. 52.—Piston Ring.

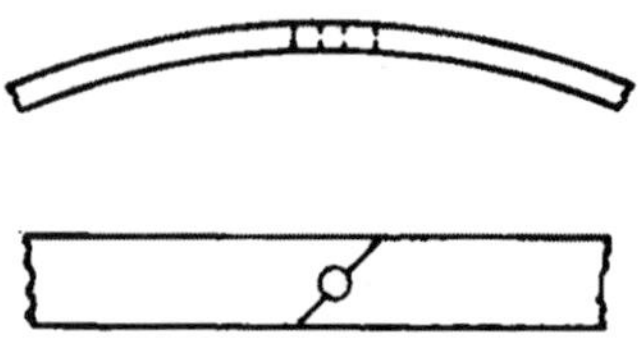

Fig. 53 —Piston Ring.

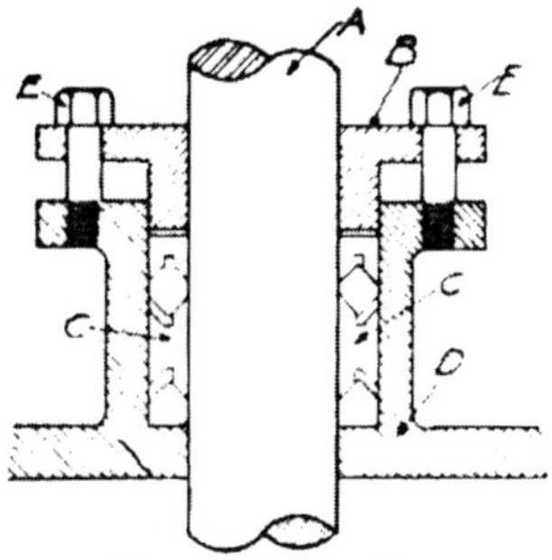

Fig. 54.—Packing Box.

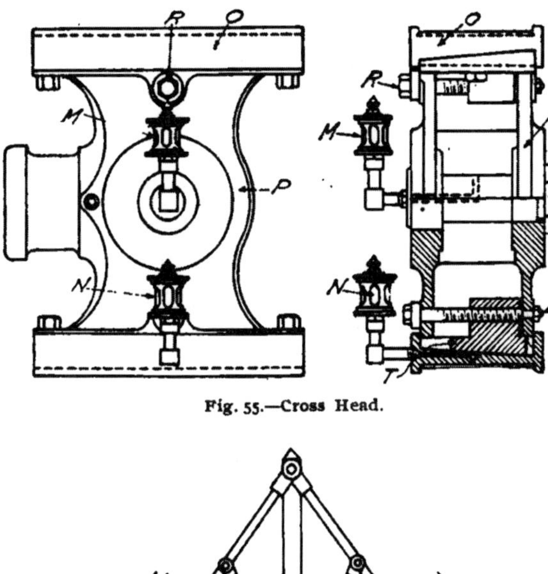

Fig. 55.—Cross Head.

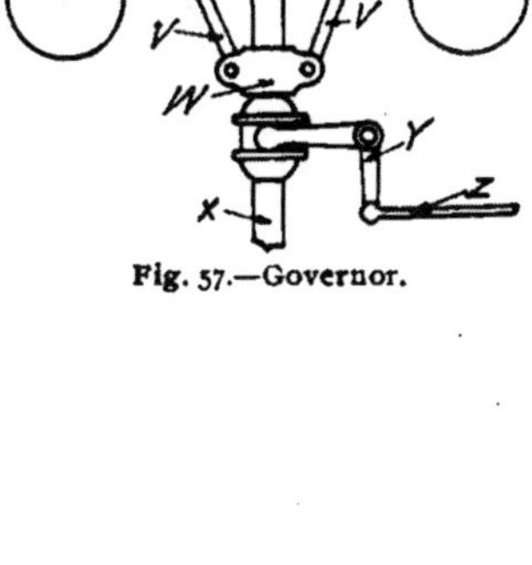

Fig. 57.—Governor.

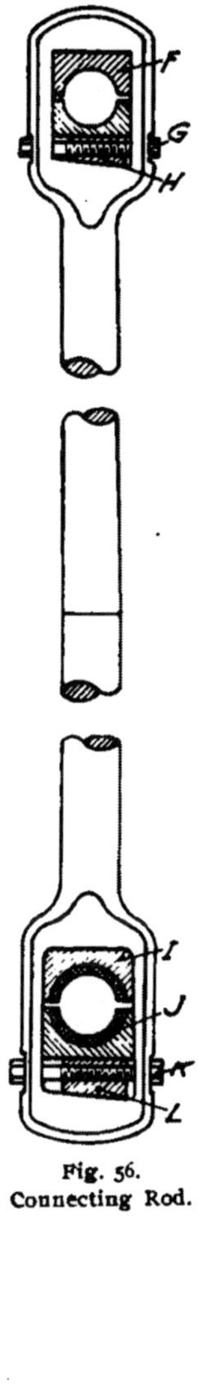

Fig. 56.
Connecting Rod.

(56) The Cross Head and Connecting Rod with pin to take up lost motion are shown in Figs. 55 and 56.

(57) The Governor.—If an engine subjected to a varying load is left to itself when the load is removed, the speed will increase and become so rapid as to result in its destruction. Watt's governor had the form shown in Fig. 57 in which the shaft is revolved. As the speed increases the balls fly out and raise the collar which cuts off steam through a series of adjustable levers YZ. Governors in which springs are used and shaft governors fastened to the flywheel have taken the place of this early, imperfect apparatus.

Many matters of great importance in connection with the steam engine cannot be treated here. It will be necessary for the student to consult special treatises for such subjects as compound engines, the Corliss trip valve and the steam turbine.

THE GAS ENGINE.[1]

(58) With the development of the automobile, the gas or vapor engine has become of commercial importance. The gas engine is a much more economical machine for many purposes than the steam engine and this has led to its introduction as rapidly as the various mechanical difficulties could be met and overcome. Within recent years large engines working on blast furnace gases have been used. A modern form of isolated power plant working on producer gas is shown in Figs. 58 and 59, with details in Figs. 60-65. The type of engine here shown is the three cylinder or marine type. Plants up to 150 horse-power and over are now of frequent occurrence and give a duty of ¾ pound of coal to 1 B. H .P.

(59) The Diesel Engine.—The familiar experiment in physics, in which a little tinder in the end of a piston barrel is ignited by rapidly compressing and so heating the contained air to the ignition point, is the central idea of the Diesel engine. The pressure may be as high as 900 or as low as 300 pounds. The fuel may be almost any inflammable liquid which is injected as a fine

[1] See Hutton, The Gas Engine, and Carpenter & Diedrich's Internal Combustion Engines.

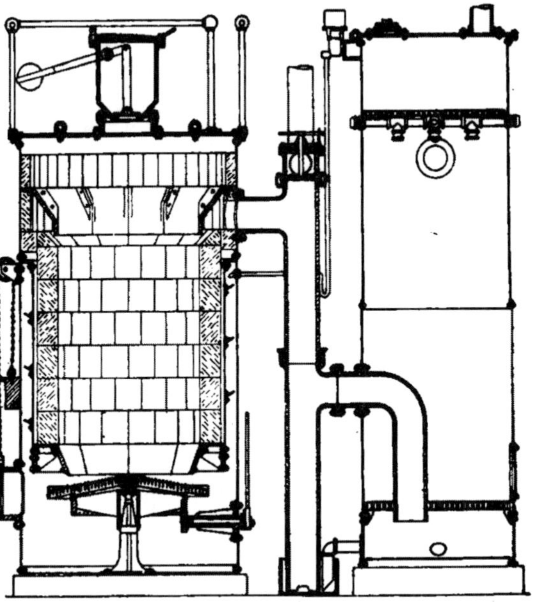

Fig. 58.—Reeves Producer.

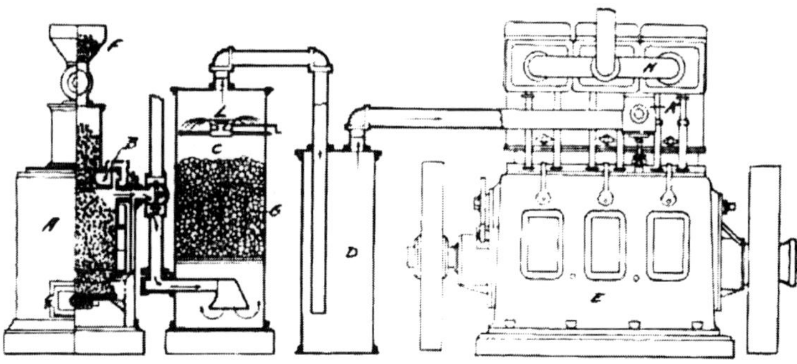

Fig. 59.—Sectional Elevation of Suction Gas Producer and Gas Engine.

A. Gas Generator.
B. Evaporator.
C. Scrubber.
D. Receiver.
E. Gas Engine.
F. Automatic Continuous Feed.
G. Coke resting on a Grating.
H. Exhaust Manifold.
K. Mixing Chamber.—Air and Gas.
L. Water Spray.

GAS ENGINE.

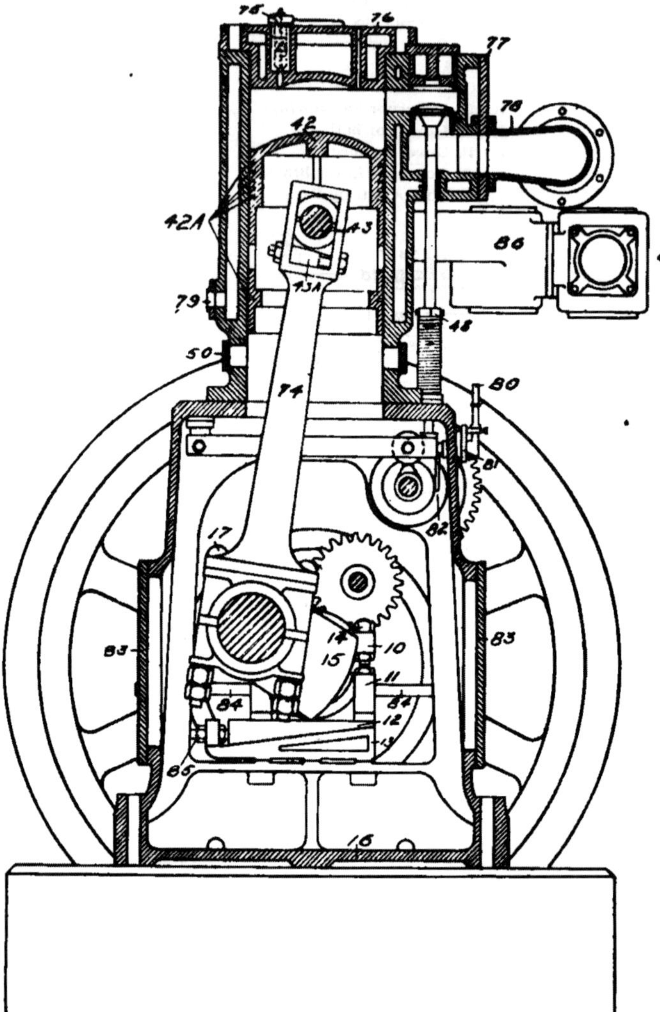

Fig. 61.—Vertical Section of Gas Engine.

42A. Piston Bushing.
43A. Cross Head Wedge.
75. Relief Valve.
76. Cylinder Head.
77. Cylinder Cover.
78. Exhaust Manifold.
79. Water Entrance Plate.
80. Relief Cam Lever.
81. Relief Cam Lever Quadrant.
82. Relief Cam Shifter.
83. Subbase Cover.
84. Center Bearing Adjusting Bolt.
85. Center Bearing Wedge Bolt.
86. Gas Manifold.
87. Mixing Valve Chamber.

24. **Cam Shaft.**
34K. Igniter Gear.
71. Governor Body.
71A. Governor Shaft.
71B. Governor Gear.
71C. Governor Shaft Bearing.
71D. Governor Ball Springs.
71E. Governor Ball.
71F. Governor Sleeve.
71G. Governor Lever Bracket.
71H. Compensating Bar.
71K. **Governor Lever.**
71L. **Regulating Spring.**
77. **Cylinder Cover.**
86. **Gas Manifold.**
86A. **Manifold Cover.**
86B. **Governor Valve.**
86C. **Valve Stem.**
87. **Mixing Chamber.**
87A. **Air and Gas Flange.**
87B. **Mixing Valve.**
87C. **Valve Lever.**

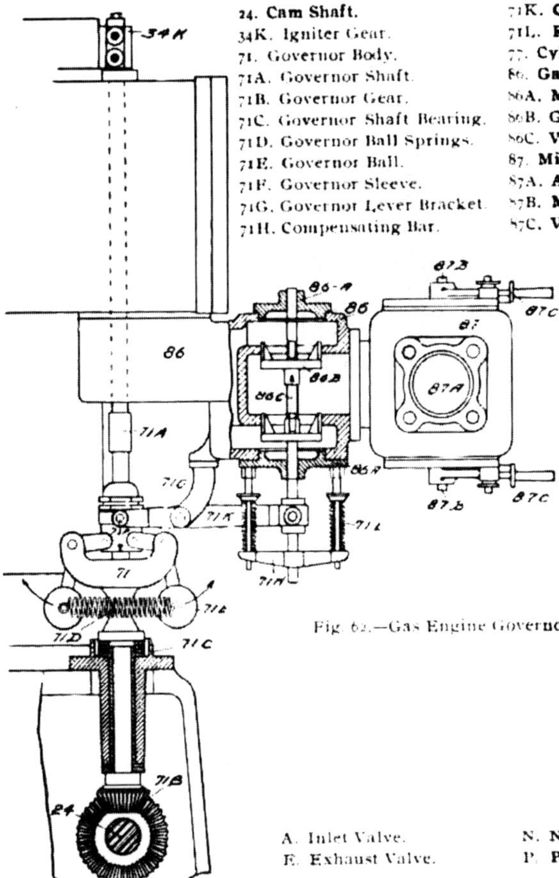

Fig. 62.—Gas Engine Governor **Mechanism.**

A. Inlet Valve.
E. Exhaust Valve.
N. **Negative Electrode.**
P. **Positive Electrode.**

Admission **Compression** **Ignition** **Expulsion**

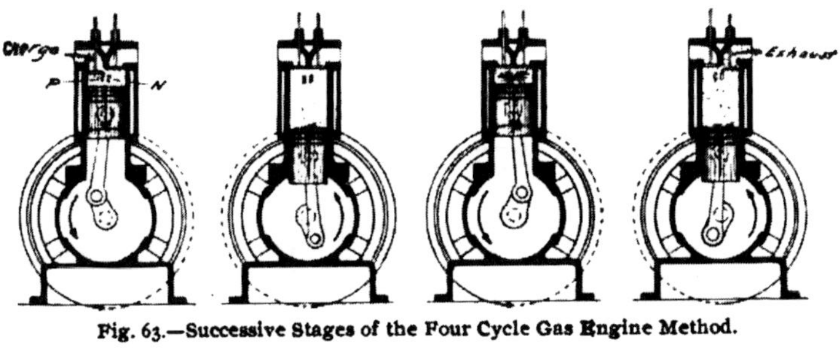

Fig. 63.—Successive Stages of the Four Cycle Gas Engine Method.

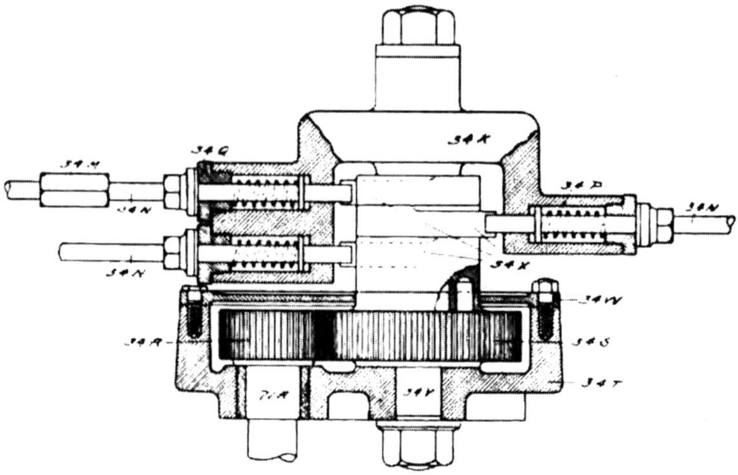

Fig. 64.—Igniter Gear for Three Cylinder Four Cycle Gas Engine.

34K. Igniter Gear Body.
34M. Extension Coupling.
34N. Igniter Trip Rod.
34P. Trip Rod Spring.
34Q. Trip Rod Guide Nut.
34R. Igniter Pinion.
34S. Igniter Gear.
34T. Gear Housing.
34V. Gear Pin.
34W. Gear Cover.
34X. Trip Rod Cams.
71A. Governor Shaft.

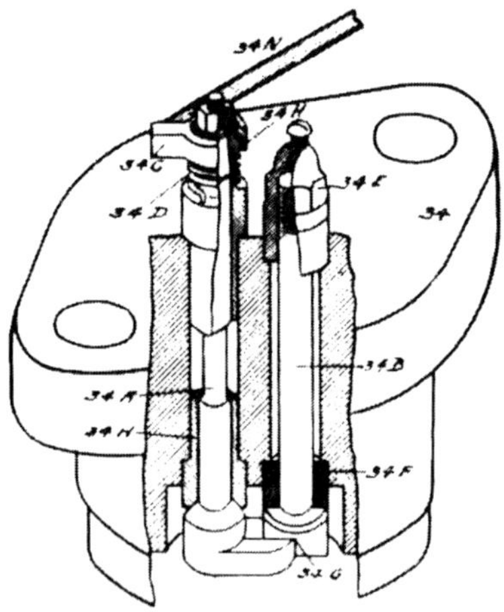

Fig. 65.—Igniter Body.—Make and Break Igniter.

34. Igniter Body.
34A. Moveable Electrode.
34B. Stationary Electrode.
34C. Igniter Pawl.
34D. Electrode Spring.
34E. Electrode Nut.
34F. Insulation.
34G. Point of Contact.
34H. Bushings.
34N. Igniter Trip Rod.

spray at the proper moment. With the heavy oils commonly used, the ignition is followed by a slow explosion which imparts a much less violent blow to the piston than in a gasoline engine. The cost of operating, with crude petroleum and gasoline at present prices, is about one-third that of a gasoline engine. The first cost of such engines is very heavy because of the number of parts, the exact fitting necessary and the high pressures to be overcome. Quite recently small engines of this type, moderately priced, have been sold and efforts are now being made to so simplify the engine that the cost may be reduced to a more moderate figure.

CHAPTER V.

PLUMBING.

The ordinary fittings for steel and wrought iron pipe are shown on Fig. 66. In fitting such pipes for carrying chemical solutions, elbows and couplings should be made as nearly as possible of the same material as the pipe line, otherwise corrosion will occur. The tools for cutting and threading pipe are shown in Figs. 67-72.

(60) Lead Pipe is frequently used for conveying liquids and gases; acid tanks are lined with lead for acid storage. The pipe lengths are connected by wipe joints if of small diameter. If of large diameter the sheets of lead are melted with a hand blowpipe fed with hydrogen and air under pressure. Where vertical joints must be made, as in large tanks, the lead burner begins by applying the flame at the bottom and proceeding upward. The surface of such a seam consists of small droplets of melted and solidified lead running the length of the seam. Lead burners acquire great skill and command high wages. Because of this and the heavy first cost of the metal, lead work is expensive. Sheet lead is quoted as 3 lb., 10 lb., etc., = lbs. per sq. ft.

(61) Stoneware Pipe made of salt glazed stoneware is frequently used. Socket joints are commonly used, or rubber connections, or clamps. The joints are best made of asbestos cord above which duro cement, to be had of the Harbison Walker Co., or some similar cement is used. This cement appears to consist of a mixture of silica, barytes and silicate of soda finely ground. It is "set" by washing with diluted sulfuric acid. See Fig. 73.

(62) Silica Tubing is sometimes used for conveying liquids. It is very durable, stands rapid changes of temperature and may be had up to 12 inches in diameter.

PIPE FITTINGS

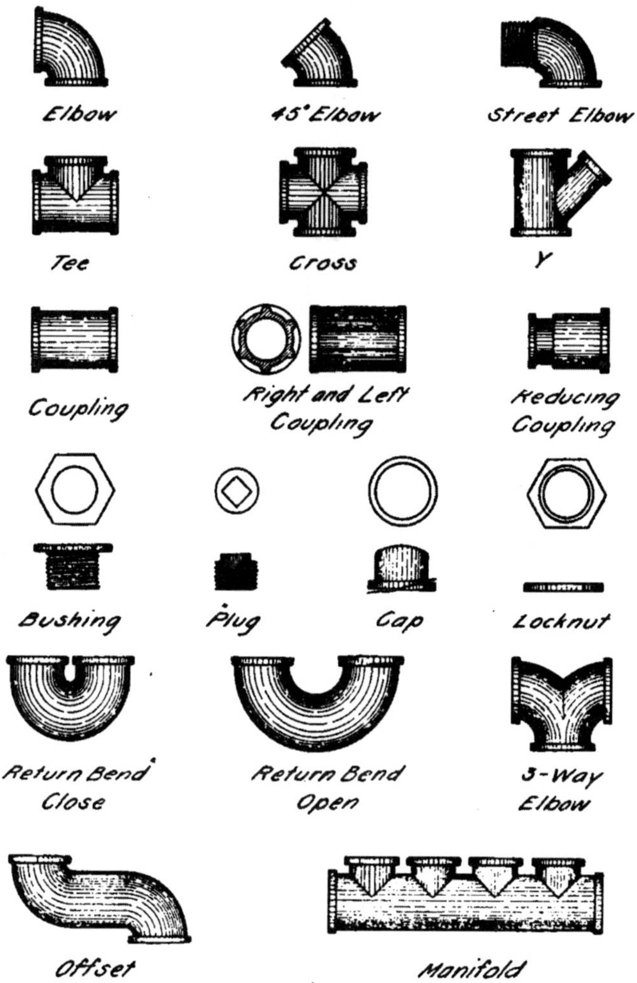

American Standard Pipe										
Size in Ins	1/8	1/4	3/8	1/2	3/4	1	1 1/4	1 1/2	2	2 1/2
Th'ds per In.	21	18	18	14	14	11 1/2	11 1/2	11 1/2	11 1/2	8
Size in Ins	3	3 1/2	4	4 1/2	5	6	7	8	9	10
Thd's per In	8	8	8	8	8	8	8	8	8	8

Fig. 66.

Fig. 67.—Tap Wrench

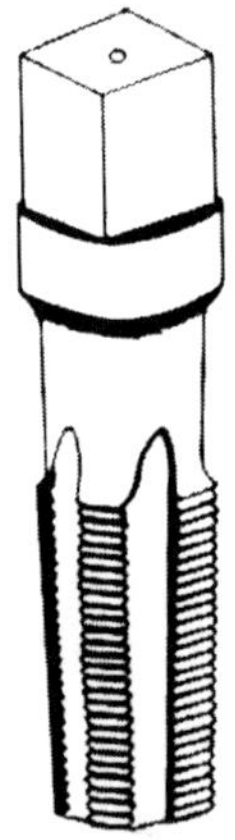

Fig. 68.—Pipe Tap.

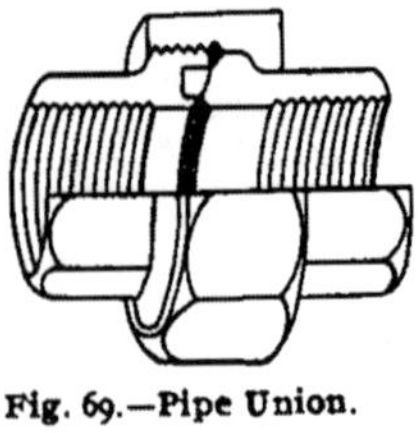

Fig. 69.—Pipe Union.

Fig. 70.—Pipe Cutter.

Fig. 71.—Pipe Threader.

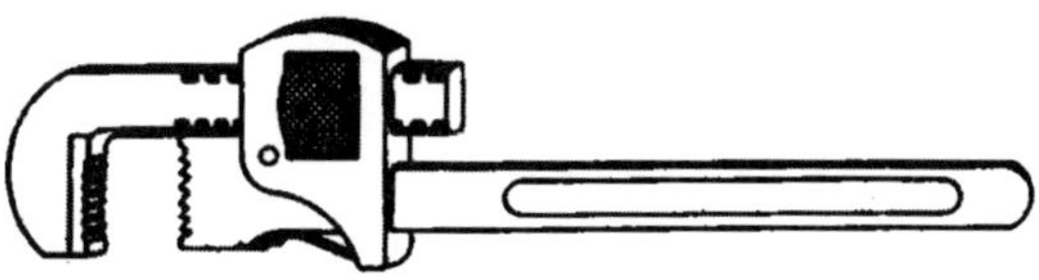

Fig. 72.—Pipe Wrench.

STONEWARE LINING.—For Use in Copper Mantles.

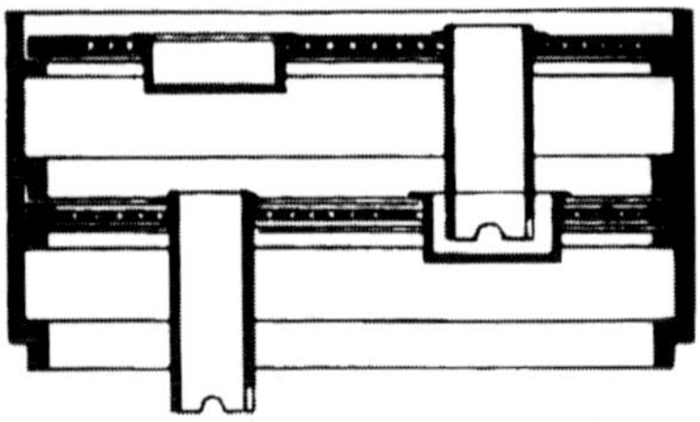

	Inches I.	II.	III.	IV.
Outside Diameter of Supporting Rings	6¾	16	17	21¼
Diameter of Sieve	5¾	15	16	20
Height of Supporting Rings	4¾	4¾	4¼	4¼

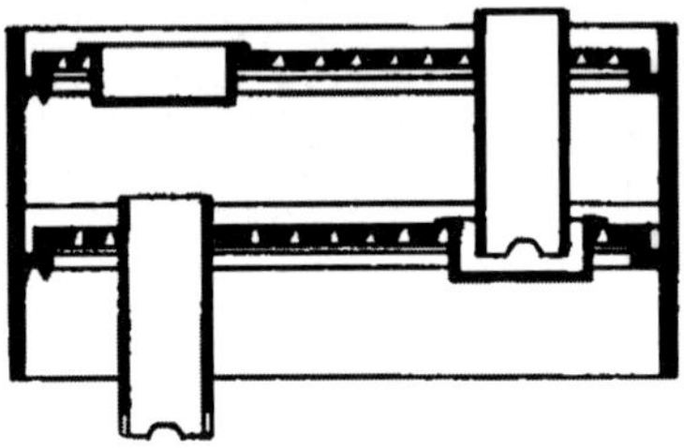

	Inches I.	II.	III.
Outside Diameter of Supporting Rings	16¼	17¾	19¼
Diameter of Sieve	14¾	16¼	17¾
Height of Supporting Rings	4¾	4¼	4¾

STONEWARE PIPE CONNECTIONS.

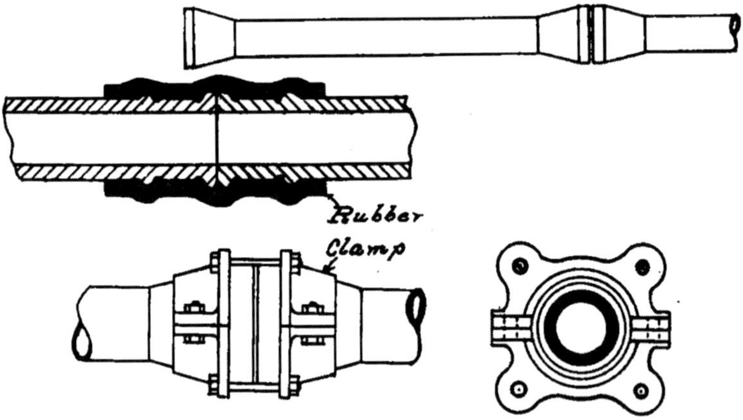

STONEWARE STOPCOCK.

Inside Diameter—Inches	⅜	¾	1	1¼	1½	2	2⅜	2¾	3⅛	3½	4	4¾

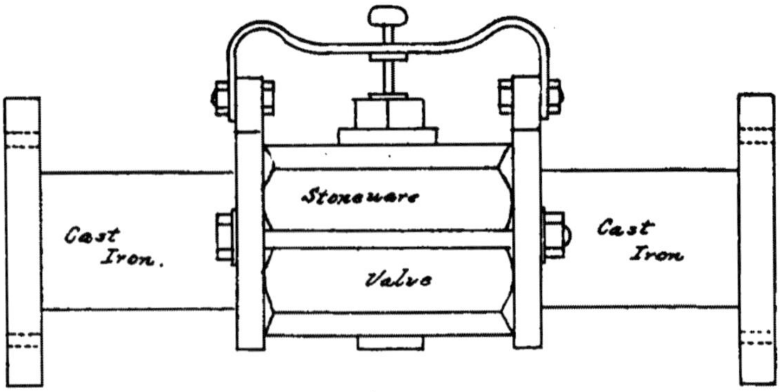

Fig. 73.

CIRCULAR STONEWARE TOWERS.
Lunge-Rohrmann-System

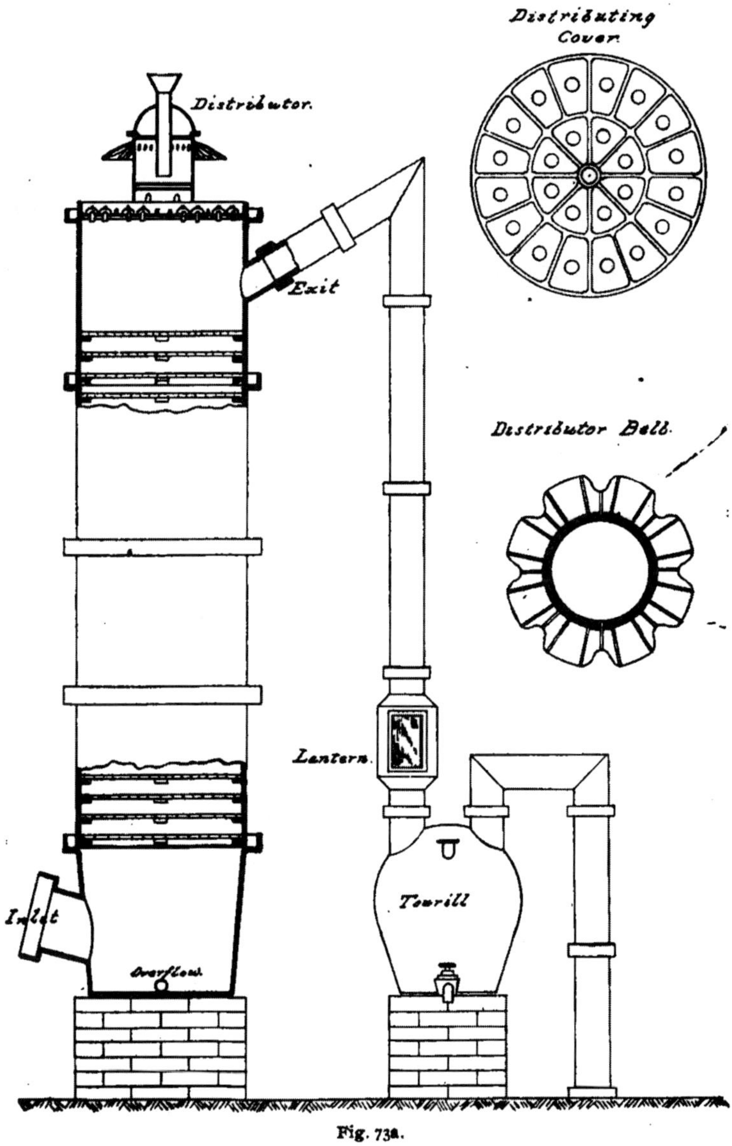

Fig. 73a.

CHAPTER VI.

CRUSHING.

(63) The coarse crushing of hard materials, such as quartz, limestone, feldspar, barytes, fluorspar, etc., is usually effected by means of some form of so-called rock breaker or crusher. The material to be crushed must, of course, be of such size that it can pass between the jaws of the crusher. If very large amounts of material are to be crushed it will pay to build very large crushers, so that large pieces may be taken into the jaws, but if smaller quantities only are needed, it will sometimes pay best to give the preliminary crushing by means of sledges. It is quite impossible to effect complete grinding of such material in one operation. The material must first be coarsely crushed. It should then be finely crushed and then ground to the size of powder wanted.

In all grinding problems it is first necessary to know exactly what is to be done, in order that an economical layout may be arranged. If only a few tons a day are to be crushed, the arrangement will be vastly different from that to be employed if many hundreds or thousands of tons are wanted. Where large capacity is needed, an expensive installation pays large dividends.

The wear and tear caused by materials being crushed vary greatly with their nature, but in general the wear on crushers is very heavy.

The substances to be crushed may be classed as follows, according to their hardness:

1. Soapstone and talc.
2. Rock salt, gypsum, graphite, soft coal.
3. Calcite, burnt lime, marble, chalk, barytes, soft limestone.
4. Fluorite, magnetite, soft phosphate, dolomite.
5. Apatite, hard phosphate, hard limestone, chromite.
6. Feldspar, magnetite, hornblende.
7. Quartz, granite, sandstone.

The machines used for this crushing may be divided as follows: First, the dry grinding, consisting of crushers, rolls and tube mills; second, the wet grinding, consisting of stamps, ball and tube mills.

(64) "Jaw crushers are built in a number of designs, but in all standard machines the jaw action or motion is identical, although there are several ways of transmitting this motion to the crushing members. The Blake design uses the Pitman construction which is probably the oldest mechanism and is a good machine to-day. During the last fifteen years, however, cam and roll crushers have entered the market and are rapidly replacing the Blake type."

(65) The Blake Crusher.—This crusher has an eccentric shaft with bearing surfaces above and below which are adjustable to take up wear. The motion is transmitted to the swing or moving jaws by toggles. The swing jaw is pivoted at the top; therefore the greatest motion of the jaws is at the discharge or bottom opening. This design has proven good but the difficulty with inexperienced or cheap labor is in keeping the eccentric boxes tight, for, if allowed to get loose and pound, this bearing soon begins to heat and to cause trouble, especially if the jaws are set too close and undue pressure is brought to bear upon these parts. This part is sometimes water jacketed to prevent heating. The friction of this bearing requires more power than is actually needed to break the rock. The cam and roll design was built to obviate these faults.

"In all standard coarse crusher designs the greatest motion of the jaws is at the bottom, ranging from 5/8 inch to 3/4 inch, according to the size of machine and work to be accomplished. This motion becomes less and less as the top jaw is approached, and when the receiving opening is reached, it is reduced to about 1/8 inch. Thus the greatest power and leverages are brought to bear at the first nip on the largest piece of rock going to the machines which fractures and breaks it partially, allowing it to slip further down by gravity between the jaws where it is again nipped, and again and again with increasing strokes until by this gradual reduction process it reaches the discharge opening where it escapes, the discharge being free and rapid on account of the maximum motion at this point. With the small motion at the receiving opening these crushers have much larger capacity than at the discharge, for at the top the rock is cracked allowing the greater part of the work to be accomplished as the material approaches the discharge. It is a great mistake to undertake to crush to less than 2 inch size in a coarse crusher. If this is attempted, packing results which clogs the crusher and puts excessive pressure on the various parts of the machine." The original Blake crusher was of cast iron. Plate steel is now often used, and Figs. 74 and 75 show how it is put together.

CONSTRUCTION OF CRUSHER.

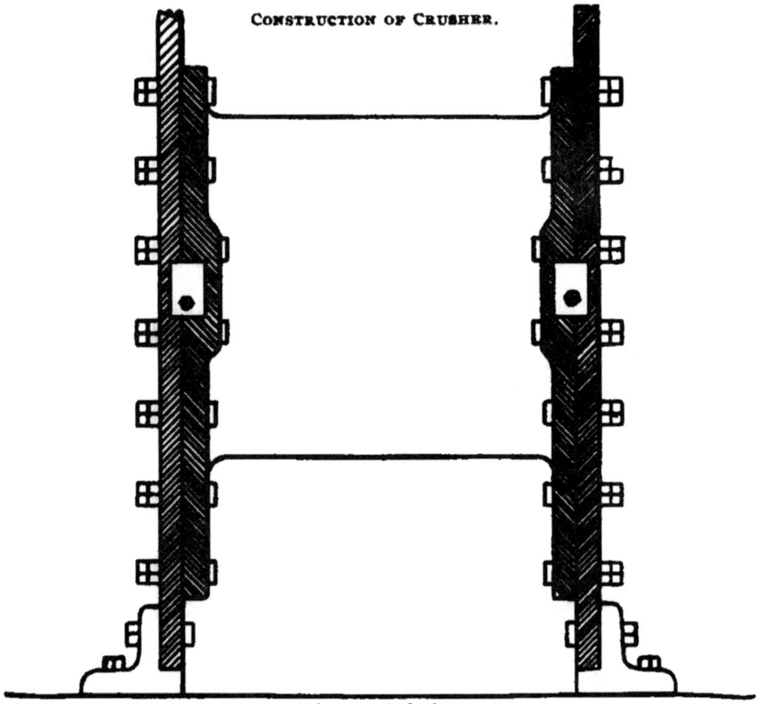

Fig. 74.—End View.

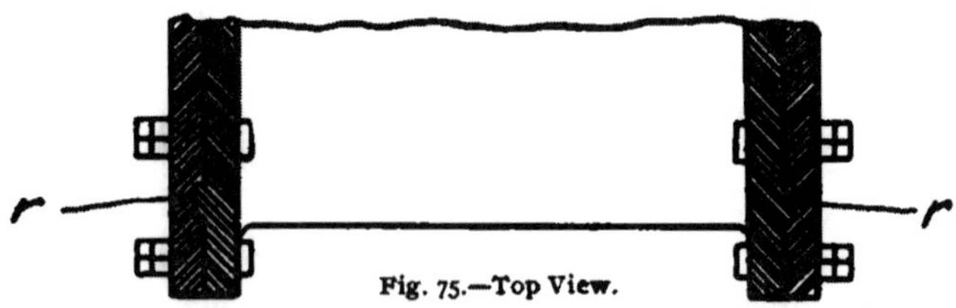

Fig. 75.—Top View.

Fig. 74. is the end view showing the method of fastening the outer and inner casing together. The bolts are relieved from undue strain by means of the rabbeting rr.

PLATE STEEL ROCK AND ORE CRUSHERS.

Type.—Jaw crusher.

Action.—Double Cam and Roll giving jaws two nips to each revolution of flywheels. Run at half the speed of others for same output. Slow speed means no hot boxes or bearing troubles and smooth steady running,—cheap foundations—long life.

Construction.—Flawless almost unbreakable Plate Steel. The side plates are securely rabbitted to front head casting and strongly tied by heavy back brace. No shearing action to bolts which simply hold parts together. Plate steel has four times the strength of cast iron and plate steel crushers weigh only one half as much as cast iron breakers.

Range of Output.—6 inches to 1½ inches.

Uses.—For crushing anything crushable that is friable.

Sizes and Capacities.—See specifications, next plate.

PLATE STEEL ROLL JAW FINE CRUSHERS.

Type.—Jaw crusher.

Action.—Same as description above of coarse breakers.

Construction.—Same as description above.

Range of Output.—1½ inches to ½ inch.

Uses.—For crushing anything crushable that is friable. These machines are for fine crushing and do the work of a coarse breaker and rolls combined, for moderate outputs. Excellent preliminary crushers to prepare hard materials for further reduction in rolls or grinding mills. Many used for concrete, top dressing for roads, etc., etc.

Sizes and Capacities.—See specifications below.

SPECIFICATIONS.

COARSE BREAKERS.

Size jaw opening inches	Capacity in tons per hour Jaws set apart 2 inches	Horse power approx.	Speed revolutions per minute	Pulley inches	Approximate weight net
**8 x 10	6 to 10	8 to 10	160	30 x 6	3,500 lb.
7½ x 13	8 to 12	12	170	38 x 8	6,000 lb.
9 x 15	12 to 18	15	155	48 x 9	8,500 lb.
*10 x 15	12 to 18	15	155	48 x 10	11,000 lb.
10 x 20	16 to 24	18	150	50 x 10	12,500 lb.
11 x 26	25 to 40	25	140	60 x 9½	19,000 lb.
*12 x 26	25 to 40	25	140	60 x 10	24,000 lb.

* Heavy Duty. ** Cast Steel.

INTERMEDIATE BREAKERS AND FINE CRUSHERS.

Sizes	Approx. capacity in tons at specified jaw settings				Horse power	Length	Width	Height	Approx. weight net
Receiving opening inches	Jaws set ap't	Per hour	Jaws set ap't	Per hour	Est. average	Ex-treme	Ex-treme	Ex-treme	Total approx. lb.
Lab 2 x 6	⅛"	250 lb. to 600 lb.	1"		¼	2' 3"	1' 10"	1' 8"	700
4 x 8	¼"	½ to 1	1"		2 to 3	3' 7"	3' 2"	2' 11"	2500
5 x 10	½"	1 to 2	1"	2 to 4	3 to 4	4' 10"	3' 0"	3' 4"	3500
6 x 15	½"	2 to 4	1"	4 to 8	8 to 10	6' 0"	4' 7"	3' 9"	7000
7 x 24	½"	4 to 5	1"	8 to 12	15 to 20	7' 1"	5' 6"	4' 7"	14000
6 x 30 Duplex	½"	5 to 8	1"	8 to 16	12 to 20	6' 0"	6' 0"	3' 9"	13000
6 x 45 Triple	½"	7 to 12	1"	12 to 20	30	6' 0"	7' 6"	3' 9"	18000
7 x 48 Dnplex	½"	8 to 12	1"	16 to 24	30 to 40	7' 1"	7' 10"	4' 7"	26000

GYRATORY BREAKERS.

(66) Gyratory breakers are of great capacity and large first cost. The crushing motion is obtained by a long eccentric shaft to which the crushing cones are attached, driven by gear and pinion. The case or frame is very heavy and has replaceable liners opposite the crushing head, which forms a circular receiving opening for the rock to be crushed. The shaft has a gyratory motion, causing an advancing and receding action of the crushing head nearly identical with that of the jaw crusher. These gyratory crushers are only suitable for large capacities; for smaller outputs the jaw crusher is preferred. See Fig. 78.

INTERMEDIATE CRUSHING.

(67) The Dodge Crusher may be used for intermediate crushing from 2 or 3 inch to 1 inch. The largest size, 11x15, has a capacity of five or six tons per hour to 1 inch size.

The Sturtevant Rock Smasher has the jaws so arranged that the pivot at the bottom of the moving jaw is placed well back and below the discharge point. Thus the moving jaw surface has a divided downward stroke which gives the material a rolling crush causing it to break down more easily than with strain pressure and also hastens the discharge. Therefore, it is less likely to clog with soft rock.

"It is to be noted that the output of a crusher depends upon three things: the setting of the jaws, the length of the discharge opening and the stroke. As the stroke is small on intermediate crushers and the jaw settings close, large outputs can be obtained only by increasing the size or length of the discharge opening. This has been accomplished in smashers by building them of multiple jaw construction. Therefore, a single machine having a jaw opening of 6x15 inches, in duplex form would have a jaw opening of 6x30; in triplex, 6x45, etc., and, as each jaw is crushing independently of the other, the strength of the entire machine resists the crushing strain of a single set of jaws. As the jaws crush alternately, the movements are balanced and these crushers run very steadily."

STURTEVANT PLATE STEEL JAW CRUSHERS.

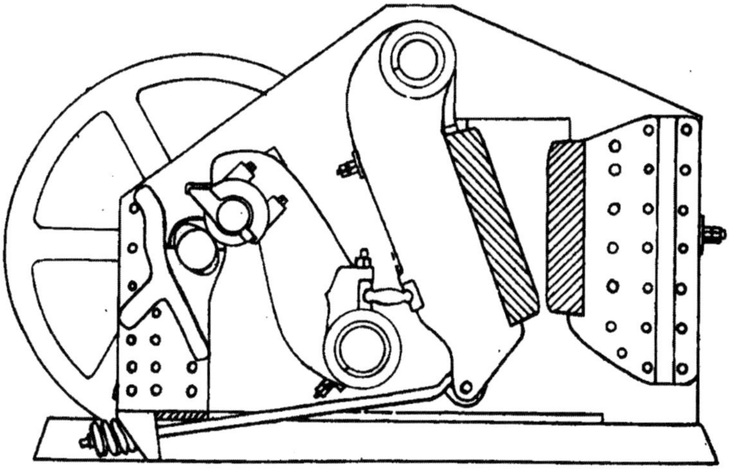

Fig. 76.—Sturtevant Coarse Breaker.

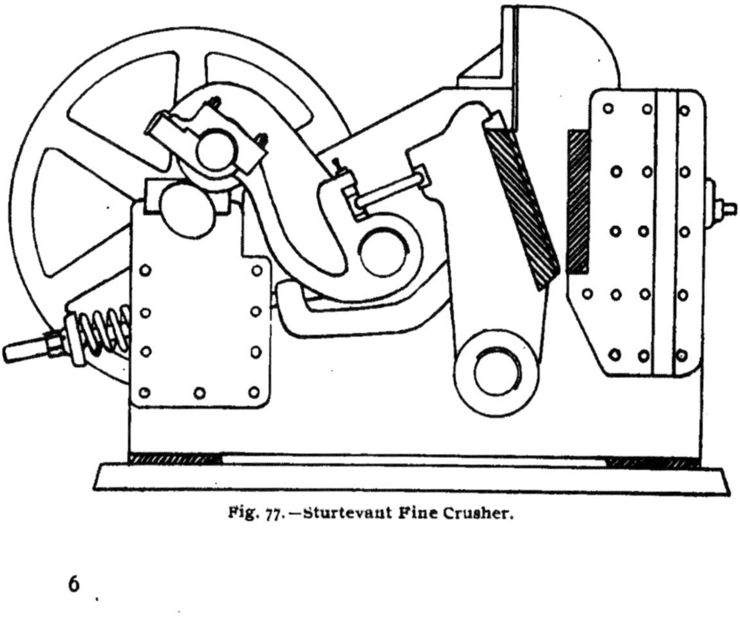

Fig. 77.—Sturtevant Fine Crusher.

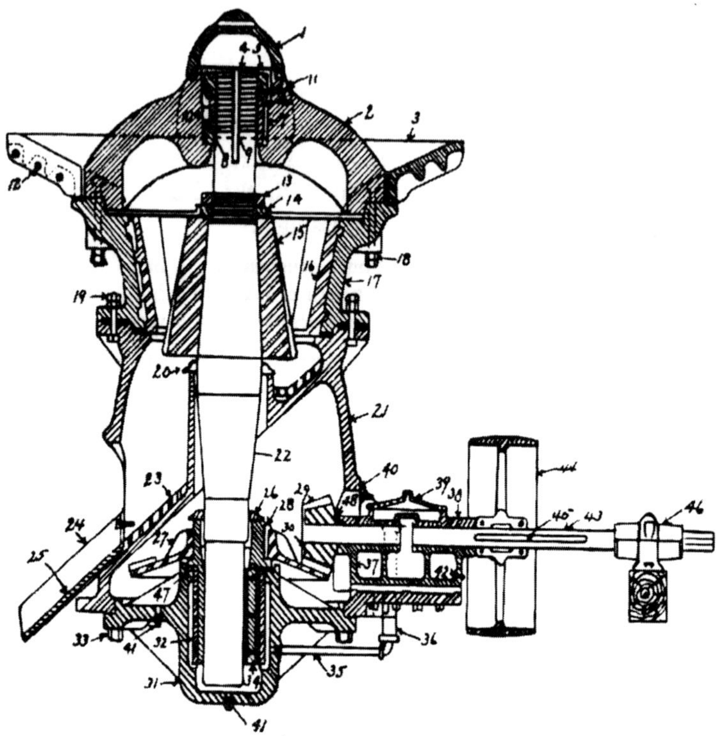

Fig. 78.—Traylor Crusher.

Sectional View.

1. Spider cap.
2. Spider.
3. Hopper.
4. Adjusting nut key.
5. Adjusting nut.
6. Sleeve.
7. Bushing.
8. Wearing ring.
9. Sleeve key.
10. Bushing key.
11. Oil feed suspension.
12. Bolts for hopper.
13. Spanner nut.
14. Bottom spanner nut.
15. Head.
16. Concaves.
17. Top shell.
18. Bolts for spider.
19. Bolts for top shell.
20. Top dust collar.
21. Lower shell.
22. Spindle,
23. Diaphragm lining.
24. Chute.
25. Lining for chute.
26. Lower dust collar.
27. Gear.
28. Key for gear.
29. Stesl pinion.
30. Key for pinion.
31. Bottom.
32. Eccentric bushing.
33. Bolts for bottom.
34. Eccentric bearing.
35. Oil feed pipe.
36. Oil reservoir.
37. Double countershaft bearing.
38. Countershaft bearing cap.
39. Oil well cover.
40. Door.
41. Drain.
42. Drain for bearing.
43. Countershaft.
44. Driving pulley.
45. Key for pulley.
46. Outboard bearing.
47. Wearing ring.
48. Washer.

(68) Figure 76 shows a Sturtevant coarse crusher of the modified Blake type. Fig. 77 shows an intermediate crusher of the Dodge type. Fig. 78 shows the gyratory crusher made by the Traylor Engineering Co., of Allentown, Pa. Figs. 74 and 75 show the construction of a crusher made of plate steel.

(69) The Power required to drive crushers varies, of course, with the material to be crushed, but with the Blake crusher is about as follows (Richards' Ore Dressing):

11-12 tons per h. p. per 24 hours crushed to 1½ inch.
15½–19 tons per h. p. per 24 hours crushed to 2 inch.
18-22 tons per h. p. per 24 hours crushed to 2½ inch.

(70) The Cost of Crushing varies with the size and character of the pieces to be crushed and the reduction accomplished from 3 to 12 cents per ton. The cost of fine grinding, as, for example, in grinding cement rock or feldspar, varies from 50 cents to $1.00 or even more per ton depending upon the fineness wanted, the amount of rejection allowed, the hardness of the material, etc.

ROLLS.

(71) Rolls are usually employed for intermediate grinding or crushing. The size of pieces which a given pair of rolls can crush depends upon the angle between the tires which must be the same as in crushers, otherwise the pieces will dance and fail to pass through. If a 36x16 inch roll is used, the maximum sized piece is 2½ inches. The maximum reduction a roll can make from this size would be three or four to one, that is, from 2½ inches to ¾ inch. The capacity of such rolls would be about 30 tons per hour. Corrugated tires are sometimes used on rolls to grip larger pieces of rock.

Rolls give less fine dust than any other method of reduction. As the above example shows, they have large capacity.

(72) Cornish Rolls have slow speed up to say 300 feet periphery speed per minute. Fast running rolls may quadruple this. The slower, the smaller the capacity for a given size. In order that tools or other unbreakable material accidentally dropped in may not destroy the rolls, heavy springs hold one set of journal boxes in place, or some other equivalent device is employed.

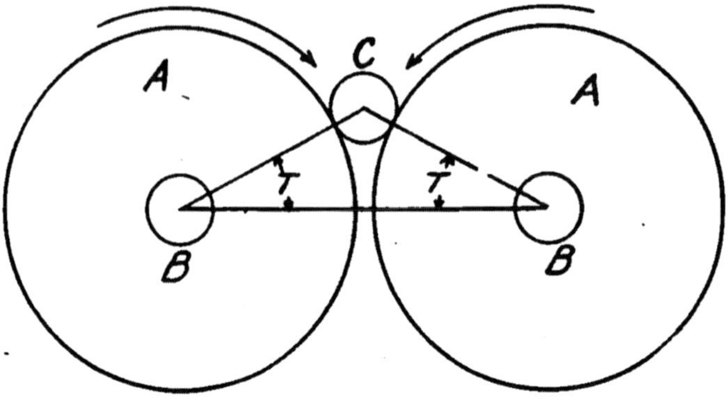

Fig. 79.

Rolls consisting of two iron cylinders, revolving on the shafts BB in the direction of the arrows and acting on the lump of ore C; one shaft in fixed, the other in movable boxes. The ore is compressed and broken and the pieces fall between the rolls.

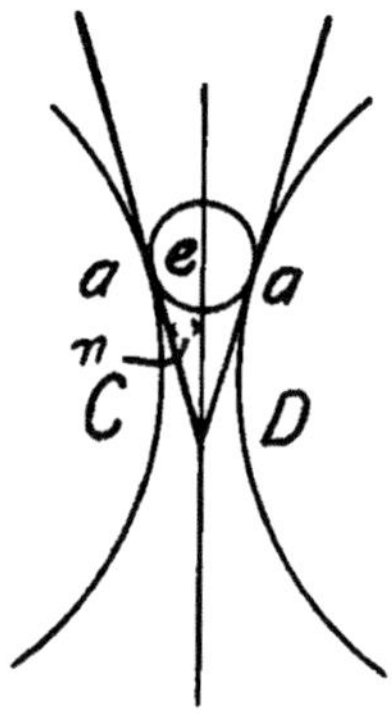

Fig. 80.

Fig. 80.—Diagram illustrating the angle of nip of rolls. The angle of nip is one-half the angle made by the tangents to the rolls at the points of contact as between the ore e and the rolls C, D; these tangents form the angle 2n, half of which is called the angle of nip. If e is as large as the opening, n = 0, increasing until the rolls cannot nip the rock which will dance upon them as they revolve. The standard angle of nip is 16°. (Richards).

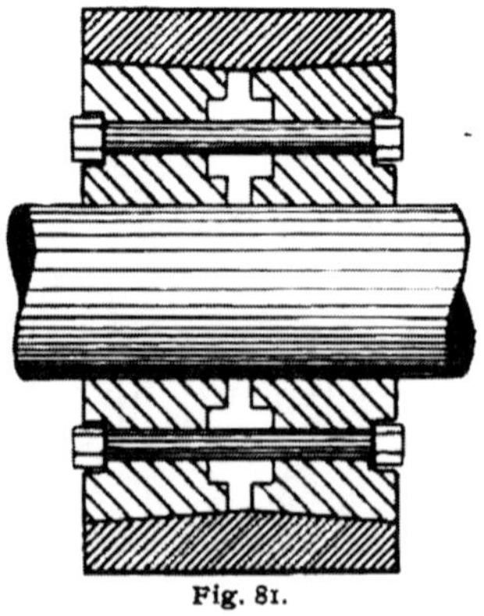

Fig. 81.

Fig. 81. — Krom's method of attaching roll shells. The core is in two parts, each a little less than half the face of the roll shells. They are slightly conical, having their lesser diameter inwards. One part is shrunk on permanently to the shaft. The other is drawn into place by four bolts.

STURTEVANT BALANCED ROLLS 21" x 10".

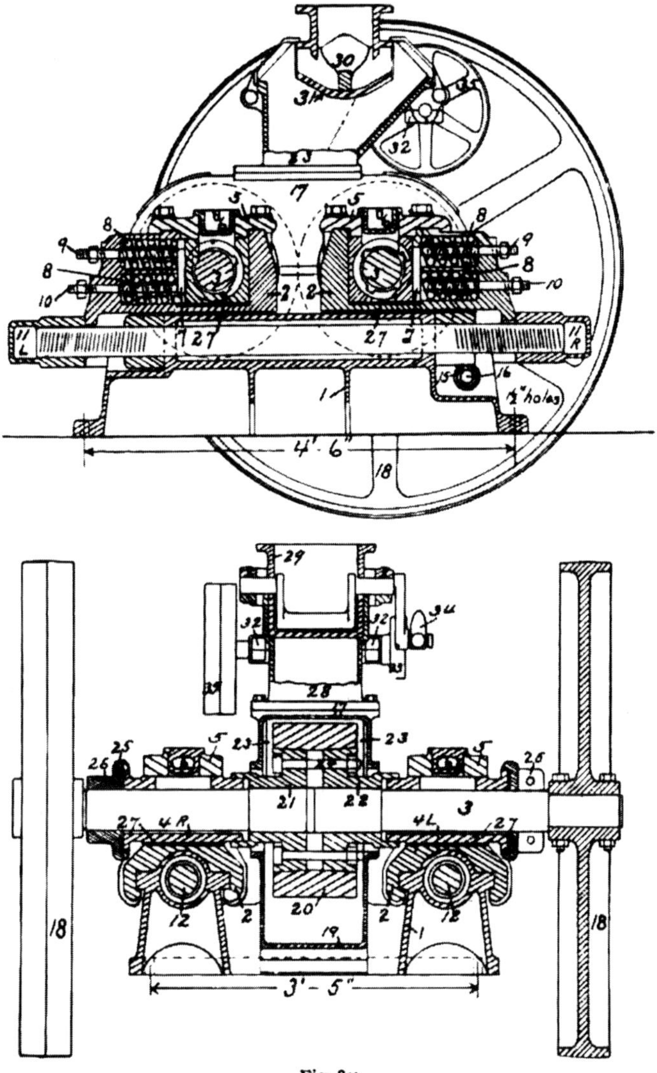

Fig. 82.

1. Bed.
2. Stand.
3. Shaft.
4. Bearing box.
5. Stand cap.
6. Bearing oil covers.
7. Spring plate.
8. Spring.
9. Upper spring stud.
10. Lower spring stud.
11. Stay bolt nut.
12. Stay bolt.
13. Stay bolt gear.
14. Stay bolt collar.
15. Worm.
16. Worm shaft.
17. Top case.
18. Flywheel.
19. Lower case.
20. Tire.
21. Solid head.
22. Split head.
23. Shield.
24. Bolts in head.
25. Shaft collar.
26. Shaft collar caps.
27. Bearing box plate.
28. Feeder base.
29. Top.
30. Sweep.
31. Trough.
32. Bearing cap.
33. Regulating block.
34. Connecting rod.
35. Pulley.

(73) Shafts are usually of mild steel; the cylinders are of soft iron in two pieces as shown in Fig. 81. The tire is made of tough hard material, such as chilled iron, cast steel, chrome or manganese steel and is held in place by the cylinders, as shown. The Taylor Iron and Steel Co., of High Bridge, N. J., furnish manganese steel tires bolted on in sections. Chilled iron costs 3-5¢ per pound, cast steel 8¢, chrome and manganese steel 12¢.

(74) Feeders are necessary to prevent clogging where no other method of regulation exists. A variety of devices are used such as rollers, pusher blocks, shakers and stirrups swinging to and fro in the ore current.

(75) The Hardinge Mill is a device for medium fine grinding. Silica lining blocks set in cement are commonly used and the ore is broken by the fall of silica pebbles, which rise up the side and drop. For medium fine grinding this apparatus, shown in Figs. 83 and 84, has been found very economical and has large capacity.

(76) The Ball Mill is likewise used for medium fine grinding and is shown in Fig. 85.

(77) The Tube Mill has a siliceous lining, is long in proportion to its diameter and is charged and discharged intermittently. It is used for very fine grinding.

(78) A Pulverizer Jar Mill suitable for laboratory is also shown in Figs. 86 and 87. This is suitable for small operations on a semi-manufacturing scale.

(79) The Progress of a Grinding Operation is tested by sifting the product through nested sieves down to 200 meshes to the linear inch and weighing the siftings. To grind material so that most of it will pass through a 100 mesh screen with, say 3 per cent. rejection, is comparatively easy. It is much more expensive to so crush that 100 per cent. shall pass such a screen.

(80) An Impact Screen for sizing large amounts made by the Colorado Iron Works Co., is shown in Figs. 88-92. This is intended for sizes between ½ inch and 80 or 100 meshes to the inch. "The motion of this screen is perpendicular, or at right angles to the plane of the screen surface, while in all flat screens the

HARDINGE CONICAL BALL MILL.

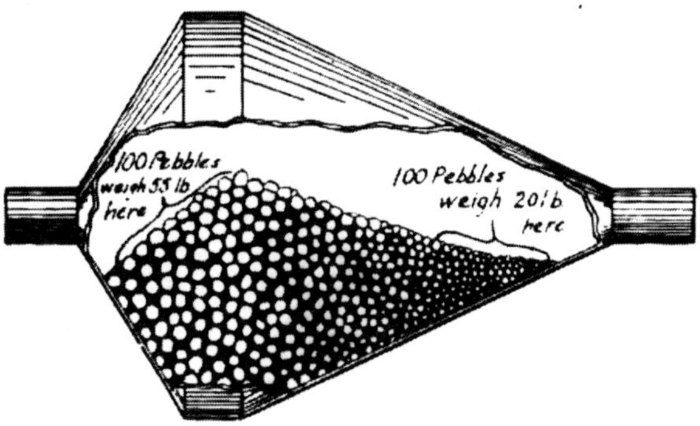

Fig. 83.—Showing distribution of balls.

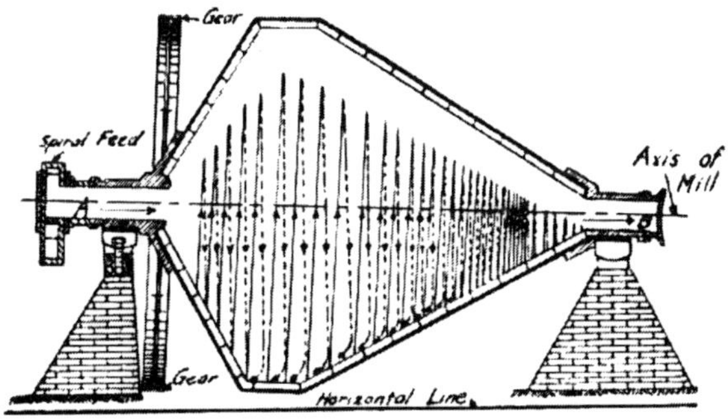

Fig. 84.—Showing path followed by balls.

The conical mill has a stepped plate lining. It is charged through A. The partly crushed material passes successively to zones of smaller balls as it is reduced in fineness until discharged through B. The forward motion thru the mill is the result of a slight inclination of the axis of the mill from the horizantal. The pebbles are of flint or round charcoal-iron slugs.

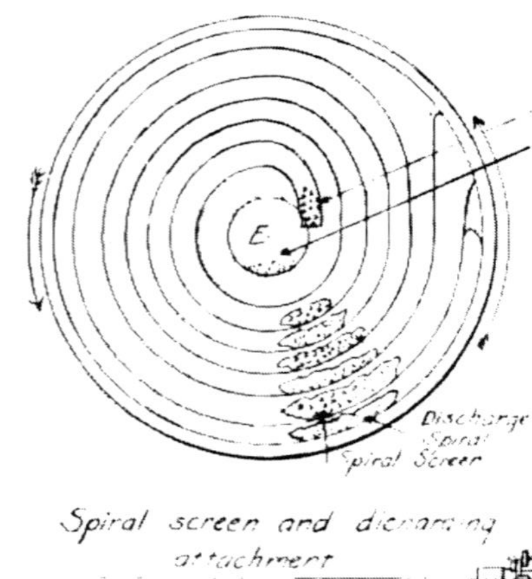

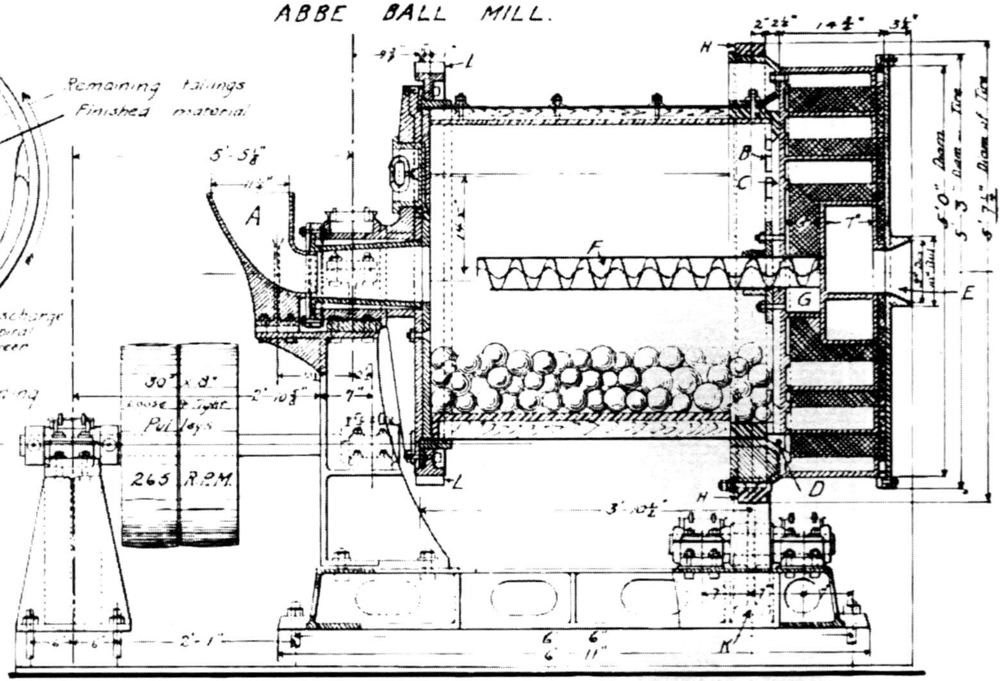

The material is fed into the mill thru A. It is then subjected to the grinding action between the balls and against the stepped lining. When it reaches the perforated plate B, the finer material passes out into a small chamber C, falls to the bottom, where, at each revolution of the mill a certain amount enters upon the screen through D. Arranged in a convolute outside of the screen is the spiral partition upon which the sifted material falls, is carried to the center and discharged through E. **The coarser material remaining upon the screen is also carried to the center, enters the tube F at G, and by means of a spiral is automatically returned to the mill. H, is the tire running on wheel bearings. K, is a wheel bearing. L, is the driving gear.**

Fig. 85.

Abbe Engineering Co., N. Y.

PULVERIZER—JAR MILL.

Jar Outside,
14¾" x 16½".

Flint Pebbles,
45 pounds.

Floor Space,
3'6" x 2'4".

Speed,
40-50 revolutions per minute.

Charge,
25 pounds.

Weight,
500 pounds.

Fig. 86.—Mill Closed.

Fig. 87.—Mill Open.

motion is nearly in the same plane in which the material is traveling. The motion of the impact screen, combined with the impact or vibration given to the screen cloth at each stroke, prevents small particles of ore from lodging in the meshes. By thus keeping the screen cloth open, its capacity is enormously greater than any form of revolving screen.

Fig. 88.—Impact Screen with Covered Housing for Dry Screening.

"A vibrating frame of wood is flexibly supported within the main frame by a pair of elliptical springs which force it against four adjustable stops. To this vibrating frame, motion is imparted by two ratchets, operating as multiple cams, keyed to a revolving shaft, upon the end of which is the driving pulley; the shaft being journaled in bearings carried on the main frame.

"For convenience in replacing the wire cloth when worn, it is fastened by means of wood strips, small wire nails or screws to a frame provided with slats placed parallel with the travel of the ore, stiffened by a cross piece one-half inch below the screen surface. This frame serves to support the cloth in a true plane. It is fitted within the vibrating frame and held in place by means of clamps with thumb nuts, permitting its removal in a few minutes without the use of tools. The screen frame and the vibrating frame which carries it act as one piece, are light in weight and constitute the only moving parts. There are no eccentrics nor reciprocating parts to cause trouble and annoyance. The screen is furnished with a steel housing, made open top for wet screening and with a wooden top for dry screening.

"When used for wet screening, a series of shallow pans is carried by the screen frame just below the surface of the wire cloth as shown, and serves to furnish the water necessary to keep the meshes open.

Fig. 89.—For Wet Screening with Distributor.

"For use in the impact screen, wire cloth is used with oblong meshes, to be so placed on the frame that the longest dimension of the openings will be in the direction of the travel of the ore, the 'size mesh' being the greatest number of meshes per lineal inch. This is to overcome the effect of the inclination of the screening surface.

"For dry screening, steel wire cloth up to and including 24 mesh, and brass wire cloth in the finer sizes are used. For wet screening, steel wire cloth up to and including 14 mesh, and brass

Fig. 90.—Screen Frame for Dry Work.

Fig. 91.—Screen Frame for Wet Work.

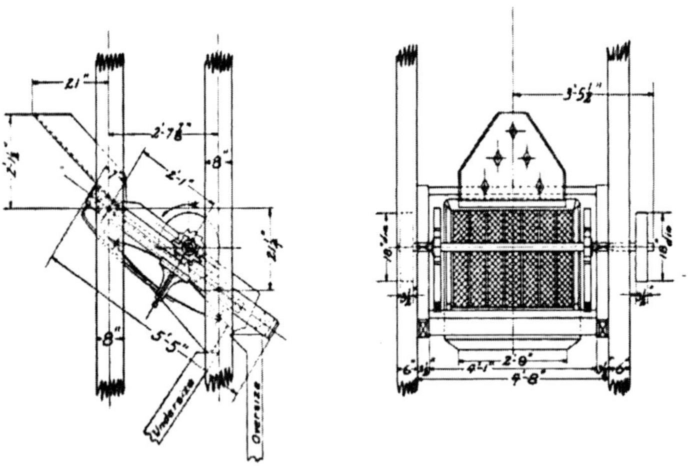

Fig. 92.—Setting for a Single 3' x 4' Impact Screen.

wire cloth in 16 mesh and finer, except under conditions where it will not answer, as where it will be subject to the action of cyanide solution are used."

(81) Other Crushing Apparatus.—The most efficient apparatus for most purposes have been described above. For special pur-

JEFFREY'S SWING HAMMER PULVERIZER.

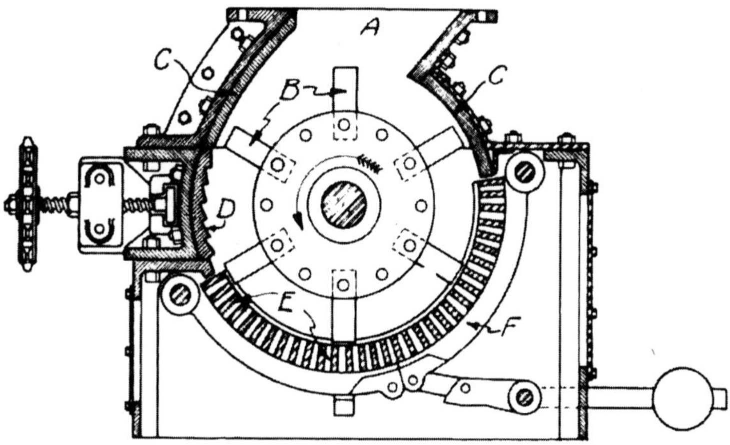

Fig. 93.—Vertical Section of Pulverizer.

A. Throat Opening into which Material is Fed.

B. Hard Steel Hammers revolving at a high rate of Speed in direction indicated.

C. Hard Iron Liners.

D. Breaker Plate adjustable to shear close to Hammers.

E. Screen Bars (any mesh) to prevent egress of oversize Material.

F. Drop Bottom to eject tramp iron or to clean out Machine.

poses Chile or Chaser mills, stamp mills, roller mills, buhr mills and beaters of a great variety of design are often employed. For coarse grinding of material like salt cake the Jeffrey Swing Hammer mill (Fig. 93), has large capacity and does the work well. Special forms of this apparatus have been designed for grinding limestone for fertilizer, glue, alfalfa, bark and a great variety of other substances.

(82) Raymond Mill.—In the Raymond Mill, Figs. 94 and 95, the substance after grinding is subjected to a blast of air which carries away the finest particles, leaving the larger, heavier pieces for further grinding. Here, also, special forms are built for each purpose, and in making enquiries, the mill builder should have full details of the material to be ground and the duty required. When practicable this method of separation of fines is preferable to screening. For beating material like dry barium carbonate into a fine powder the Mead Mill answers very well and has large capacity. Pulverulent material like this must be carefully dried or it will choke the mill.

(83) Williams Mill.—Fig. 96 shows an impact mill built by the Williams Co., St. Louis, Mo., for pulverizing bark. It has a capacity of three to four cords per hour and requires 30-35 horse-power.

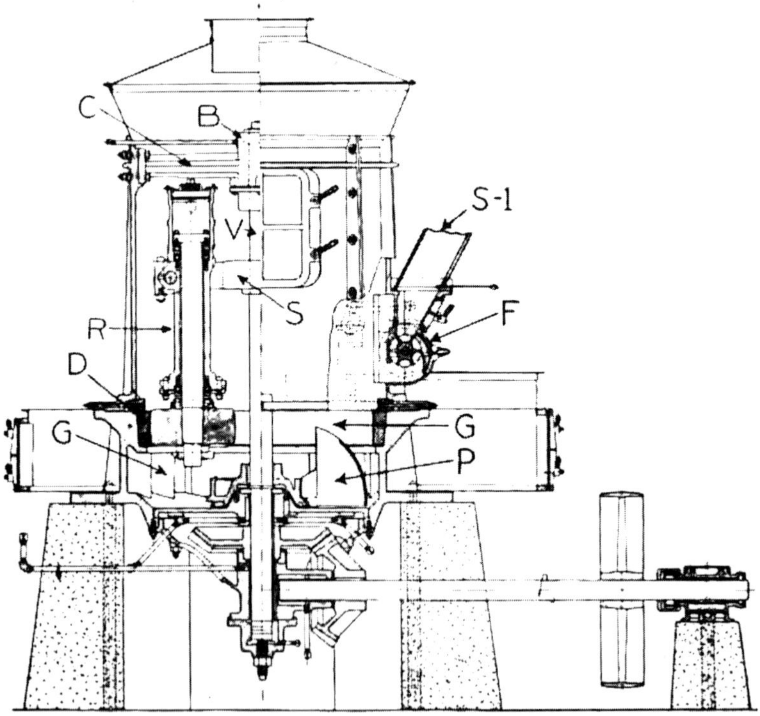

Fig. 94.—Raymond Mill.

Fig. 95.—Raymond Mill.

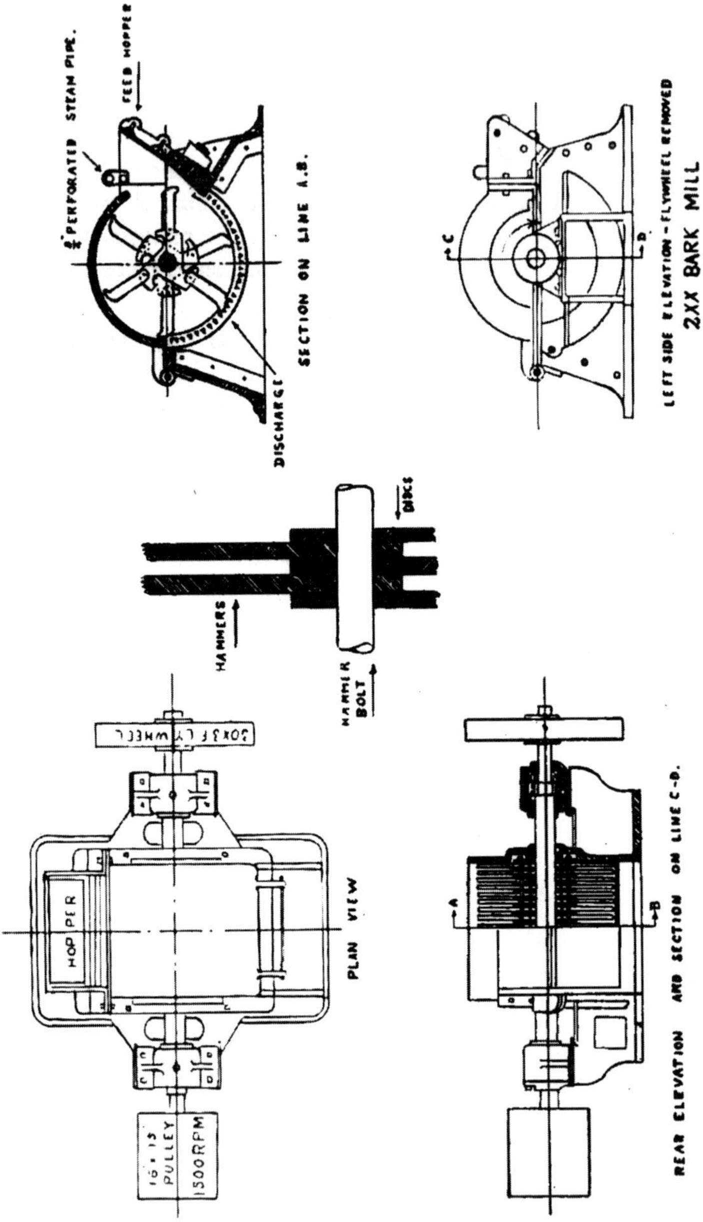

Williams Patent Crusher and Pulverizer Company, Engineering Dept. St. Louis, Mo.

Fig. 96.

CHAPTER VII.

DISSOLVING.

(84) The act of bringing substances into solution appears so simple as to require no comment. A little experience shows this to be wide of the mark. In some cases it is difficult to obtain cheaply and rapidly a saturated solution of the material wanted and, inasmuch as a partly saturated solutions means more evaporation, it is to be avoided.

(85) Salt solutions are almost always heavier than the solvent. If, then, we place the substance to be dissolved in the bottom of the tank, the heavy saturated solution will form and remain just around and over the solid, while the upper part of the tank will contain only solvent with no or only very little dissolved material. If the material to be dissolved is placed in the upper part of the tank in an open sieve or other appropriate support, as fast as the solution forms, it drops to the bottom of the tank and fresh solvent takes its place until the whole amount of liquid becomes saturated. The rapidity of solution and hence the amount of saturated solution to be made in a given time depends upon several factors.

1. Solubility: The more soluble the substance, the more rapidly solution takes place.

2. Surface exposed: The larger the surface, the more rapidly solution takes place.

3. Temperature: Usually, the higher the temperature, the more rapidly solution takes place. Solution is sometimes complicated by reactions which take place when the solute is brought in contact with the solvent. For example, dry sodium carbonate and sulphide in fine powder combine with water with a rise in temperature to form lumps which have much less surface than the original material.

(86) Stirrers are frequently employed to hasten solution. They may be made of steel, or when necessary of steel covered with lead, or entirely of wood. The John Johnson Co., of Brooklyn, N. Y., make a turbine mixer shown in Fig. 97. Solution may be hastened by coils or jackets on the bottom or around the inside of the tub, heated by open or closed steam. The material is shoveled or dumped in at the top. It is best dumped in from an automatic trip weighing scale, as in this way more uniform results are obtained.

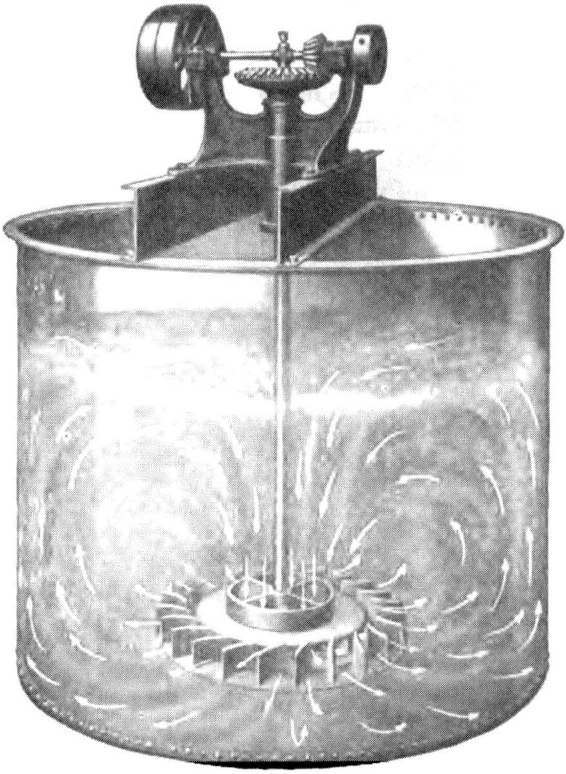

Fig. 97.

(87) A dissolver made by the Werner and Pfleiderer Co., of Saginaw, Mich., is shown in Fig. 98. This is made in varying sizes from 2.65 to 1250 gallons working capacity. The makers state that one customer has reported that they dissolved 750 pounds of borax in 9000 pounds of cold water in from 10 to 12 minutes in the largest size dissolver, 5 horse-power being required. This machine may also be used as a mixer for preparing thin pulp. As an illustration, one concern with a 500 tank dissolver using 300 gallons of lukewarm water to 1500 pounds of clay per batch, finished the operation in 1½ hours with 2½ to 3 horsepower. This includes the power required to drive a pump which pumps the mixture from the machine to the coating room where it is used for coating wall paper.

Fig. 98.—Rapid Dissolver, Size 1.

(88) Compressed Air is often used for stirring, and answers very well with Pachuca tanks (Fig. 99). In some cases, as in dissolving raw sugar, the material is dropped into the tank, water run in and open steam blown through until solution takes place.

(89) Where obstinate setting takes place, as with barium sulphide, the dry material may be fed with warm water in regulated streams through the trunnion of a ball or Hardinge mill. The rotation of the apparatus, with the grinding action, gives rapid and perfect solution. The solution also heats considerably, due partly to the heat of combination and partly to the conversion of motion into heat.

(90) The Buff-Dunlop or Shanks Apparatus.—In this apparatus fresh material meets a nearly saturated solution, after this a progressively weaker liquor and finally water is employed. This

PACHUCA MIXING TANK.

Fig. 99.

apparatus has been employed for lixiviating black ash for many years. Four tanks are used as shown in Fig. 100. Each tank is 7x7x5 feet and will hold black ash from 6000 pounds salt cake. Such a set can wash the ash from 12,000 pounds salt cake in 24 hours. The tanks are made of 5/16 to 3/8 inch boiler plate properly stiffened by angle irons. Each tank has two overflow pipes, one, *FFFF*, for strong (Fig. 101), the other, *EEEE*, for weak liquor. The weak liquor pipes allow the weak liquor to flow from one tank to the next in the series, while the strong liquor pipes empty into the strong liquor launder, *I*, emptying into the strong liquor well, *D*. Those for strong liquor have the outlet and plug placed lower than in those for weak liquor. These pipes terminate below the false, perforated bottoms of the tanks which are covered with three inches of sifted cinders to act as filters. Three

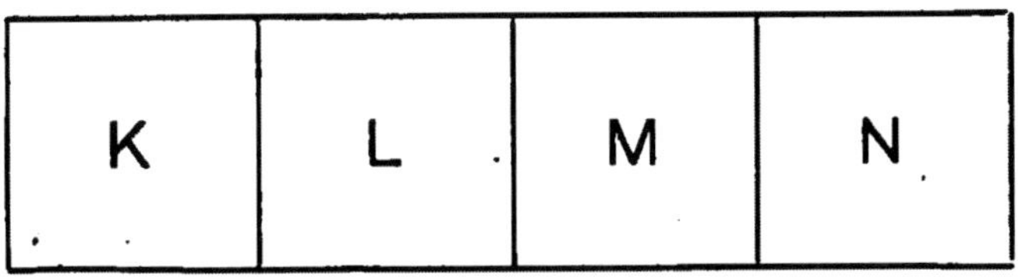

Fig. 100.

tanks, *L*, *M*, *N*, are in operation (Fig. 100), while the weak liquor in *K* is being discharged through the cock *G* into the launder *H* and weak liquor well *C*. The waste in *K* is then shoveled out and the tank charged with fresh ash. The weak liquor is pumped from the well *C* upon the surface of the nearly exhausted charge in *L* through the pipe *B*. This liquor passes down through the charge in *L*, gaining in strength continually, and up through the second pipe *E* into *M*. It flows down through the charge in *M* and up through the third pipe *E* into *N*; then down through the fresh charge in *N* and up through the fourth pipe *F*, now entirely saturated, into the launder *I* and strong liquor well *D*. By means of the pipe *J* weak liquor from *N* may be discharged upon the surface of the charge in *L*, thus making any tank first, second, or third in the series. Instead of using cocks on the ends of the strong liquor discharge pipes *F*, a short piece of pipe is screwed into the elbow. When this is turned down at a right angle the liquor discharges into the launder *I*.

BLACK ASH LIXIVIATING TANKS.

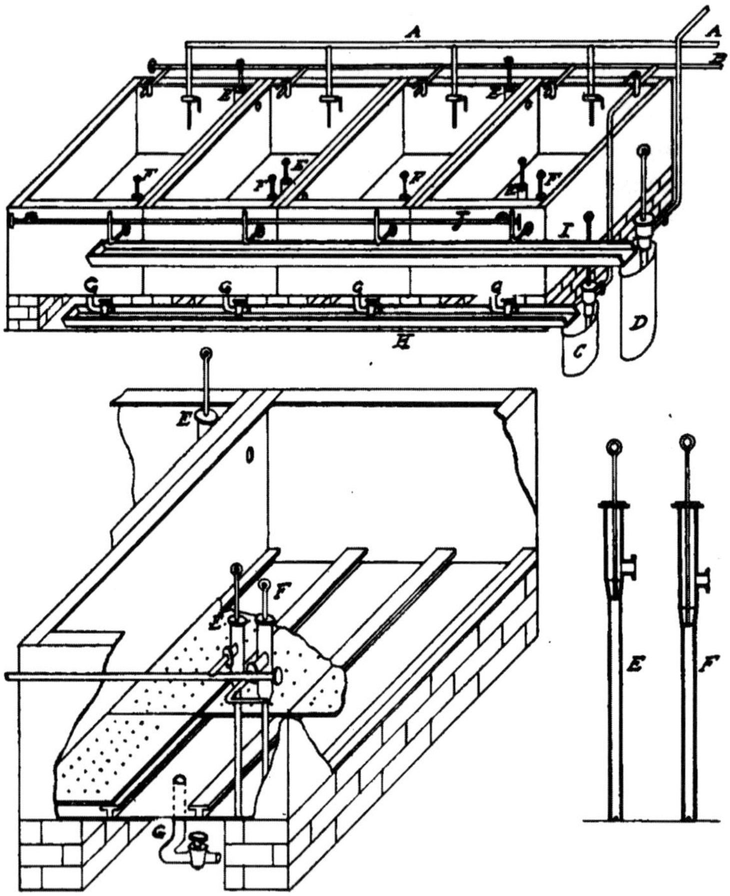

Fig. 101.

A. Steam Pipe.
B. Water or Weak Liquor Pipe.
C. Weak Liquor Well.
D. Strong Liquor Well.
E. Overflow Plugs for Weak Liquor.
F. Overflow Plugs for Strong Liquor.

CHAPTER VIII.

FILTRATION.

(91) The simplest method of filtration is that of filtration by decantation. The solution is allowed to stand in a tank, and the clear solution run off. Sometimes a bag is tied over the end of the discharge pipe to catch floating straws, etc. In such cases the bag must be removed periodically, turned inside out and carefully washed. Where color is a criterion of value, as, for example with blanc fixe, too much care to obtain bright, clear solutions cannot be taken.

In some cases clarification may be aided by adding a little albumin and heating, or by adding fine asbestos and stirring thoroughly, or by causing the formation of a precipitate, as by adding a trace of alum or calcium chloride, or some fullers earth, or precipitated silica. The nature and amount of material added must depend in each case upon the nature of the solution and must be ascertained by preliminary experiment. I cannot impress too strongly the desirability of trying earnestly to get the desired result by settling if this be possible. It is nearly always necessary to filter the last part of the solution, but filtration is always tedious and expensive.

(92) Separators.—If a stone be tied to a string and rapidly revolved by the hand and arm and the string and stone released, the stone will fly in a line tangent to the circle described at the moment of release. It may be shown that the force exerted is $F = \frac{mv^2}{r}$ where m = mass, v = velocity and r = radius of the circle described. The force necessary increases directly with the mass and velocity and decreases as the circle increases in radius. If we suppose the velocity and radius to be constant, the force necessary will vary directly with the mass.

Mass is the product of density into volume, or mass = Vd.

If we assume that V is constant it becomes evident that as the density increases or diminishes, the mass will increase or diminish, and the force required will proportionately increase or diminish. We may write this $F = \frac{Vdv^2}{r,}$, or, assuming r, V and v^2 to be unity,

$$F = d.$$

That is, as the body grows in density the centrifugal force increases. This is the principle of the cream separator (Fig. 102), in which the whole body of new milk is revolved. The fat or

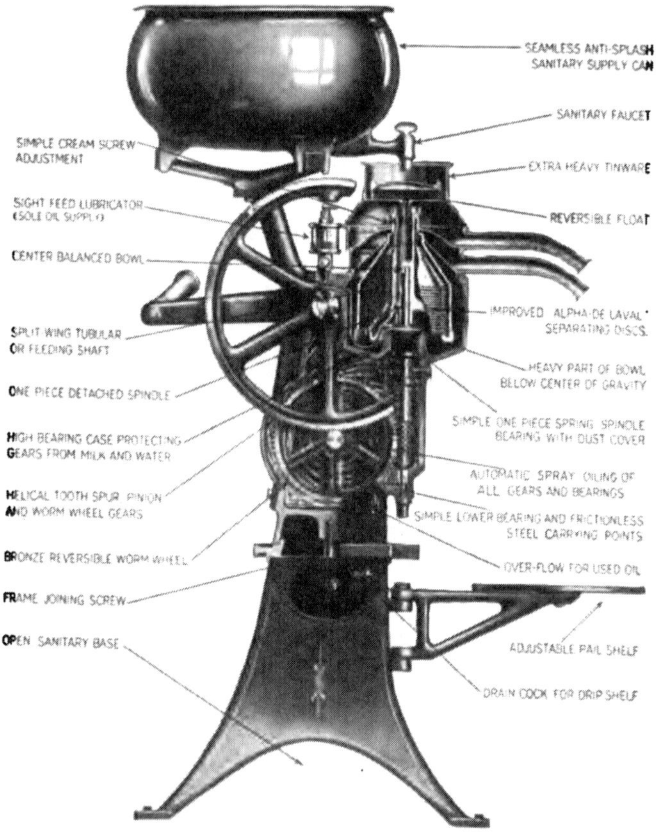

Fig. 102.—Improved De Laval Cream Separator
(From Vulté & Vanderbilt's "Food Industries," p. 180)

cream having a lower density than the milk, flows to the center, and the heavier skim milk flows toward the periphery. The apparatus is provided with an inlet tube at one end and two outlet tubes at the other. By means of proper valves the volume of each outflow may be regulated, and a cream of the desired fat content obtained.

Milk is an emulsion of two liquids and the same method may be applied to the separation of other emulsions. In these cases we have the separation of two liquids.

The same method may be applied to the separation of a solid from a liquid, as in filtration. Thus, Gore[1] has shown that cider which contains considerable solid material is best clarified by means of a cream separator. He has shown that when a separator capable of delivering 450 pounds of milk per hour is employed, 45 gallons of fresh cider may be clarified in an hour.

Three forms of apparatus constructed on this principle were described in 1911 by the De Laval Separator Co., 165 Broadway, New York City.[2]

(93) It must not be forgotten that in order to effect separation, it is necessary that there must be a difference in density. Where no difference in density exists, the liquid will remain cloudy. Thus Gore found it impossible to get a perfectly clear cider with the separator even after passing twice through the apparatus and De Laval's practically acknowledged the same thing to be true of some mixtures, the apparatus described in the second citation being a separator and filter combined.

(94) Separation is often very desirable as a preliminary to filtration, thus at the Fleischmann yeast plant at Peekskill, N. Y., the yeast after 18 hours growth in a wort is run through De Laval separators resulting in a thick paste of yeast cells and a clear filtrate. The thick paste is then run through a center feed filter press, result: a dry cake such as we find on the market.

(95) The Sharpless Super Centrifuge is shown in Fig. 103. The smaller size for laboratory use operates at 40,000 r. p. m. and the largest at 17,000 r. p. m. In this apparatus therefore the force of gravity is multiplied by these figures. The laboratory size is provided with an inner bowl shown in Fig. 104. The theory of

[1] H. C. Gore, Unfermented Apple Juice, Bull. No. 118, Bureau of Chem., U. S. Dept. of Agriculture, Page 15,

[2] *Met. & Chem. Eng.*, 9, 147, 159, 283.

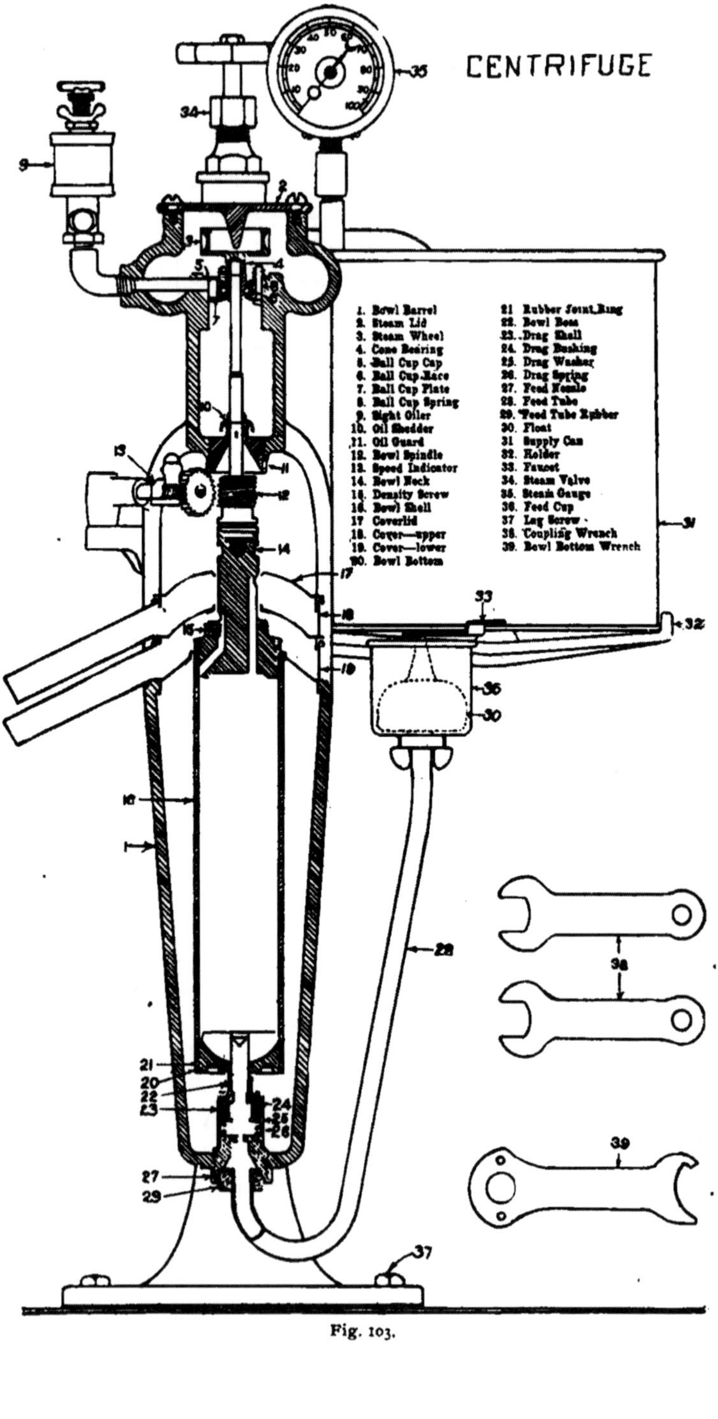

CENTRIFUGE
1. Bowl Barrel
2. Steam Lid
3. Steam Wheel
4. Cone Bearing
5. Ball Cup Cap
6. Ball Cup Race
7. Ball Cup Plate
8. Ball Cup Spring
9. Sight Oiler
10. Oil Shedder
11. Oil Guard
12. Bowl Spindle
13. Speed Indicator
14. Bowl Neck
15. Density Screw
16. Bowl Shell
17 Coverlid
18. Cover—upper
19. Cover—lower
20. Bowl Bottom
21 Rubber Joint Ring
22. Bowl Boss
23. Drag Shell
24. Drag Bushing
25. Drag Washer
26. Drag Spring
27. Feed Nozzle
28. Feed Tube
29. Feed Tube Rubber
30. Float
31 Supply Can
32. Holder
33. Faucet
34. Steam Valve
35. Steam Gauge
36. Feed Cup
37 Lag Screw
38. Coupling Wrench
39. Bowl Bottom Wrench

Fig. 103.

centrifugal separators has been considered by Eugene E. Ayers, Jr. (*J. Soc. Chem. Ind.* **35**, 676).

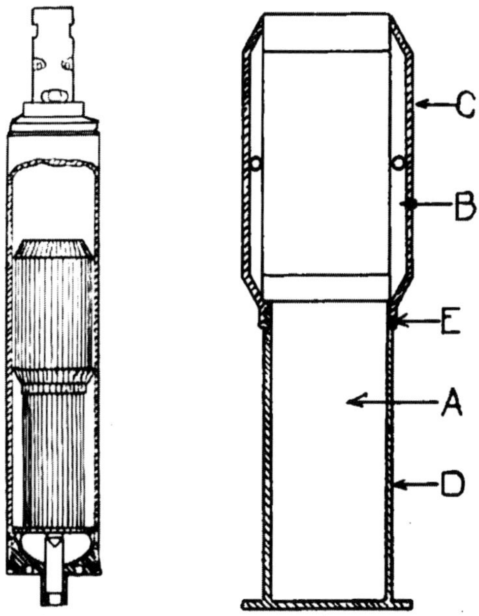

Fig. 1. Fig. 104. Fig. 2.

The Inner Bowl is a device for the clarification or filtration in batches of 50 cc. or less of a liquid. It is non-continuous. By disconnecting top (C) from the bottom (D) by means of the screw (E) the lower chamber (A) may be filled with the liquid to be clarified. The top is again screwed on and the device placed in the regular Laboratory Super-Centrifuge bowl as in Fig. 1. As the bowl starts to revolve, the liquid is almost instantly drawn into chamber (B) where clarification takes place—the solids being firmly packed by centrifugal force generated against the walls of the upper chamber. When the bowl is stopped the liquid will again collect in the chamber (A) from which it can be removed for examination. The solids can then be removed from the upper chamber. In most cases quantitative results may be secured.

This device is valuable for any clarification work where only small amounts of liquid are available, but will be appreciated in handling serum or bouillons.

CENTRIFUGALS.

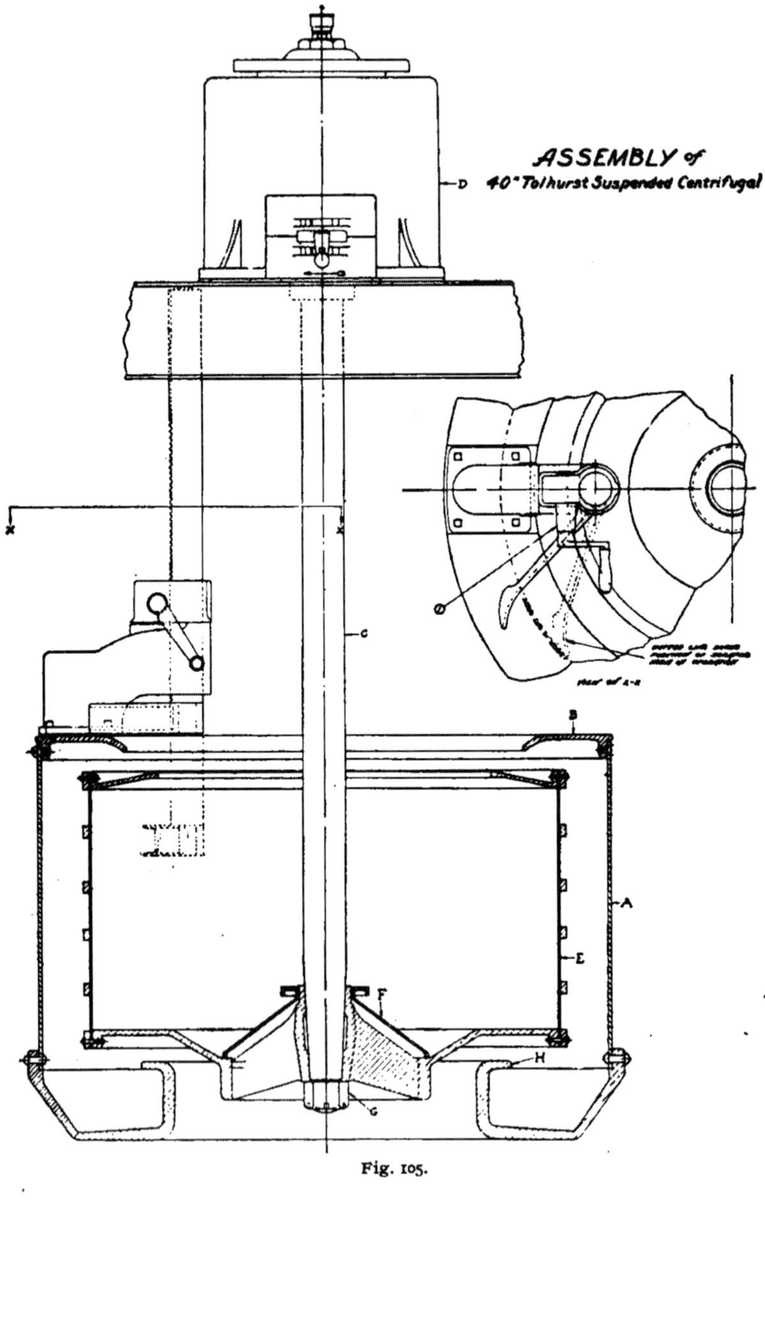

Fig. 105.

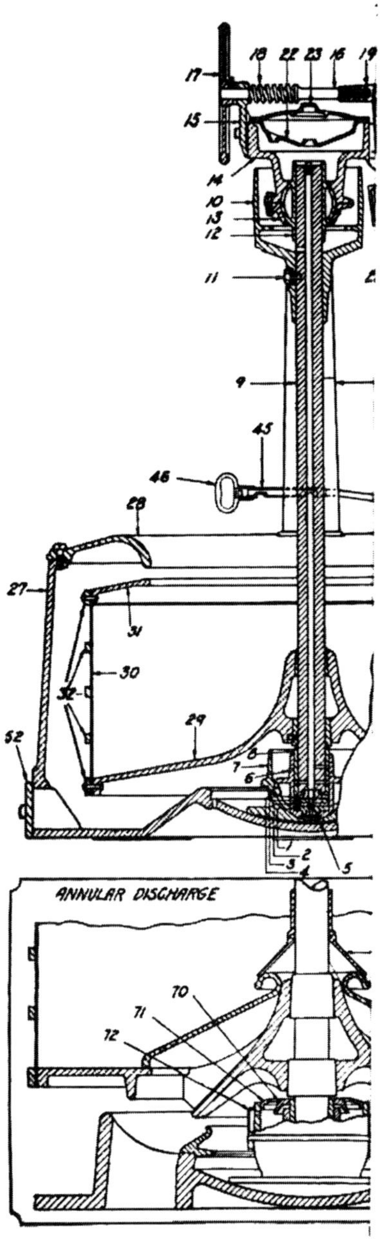
ANNULAR DISCHARGE

(96) Centrifugals are a special form of separator used for rapidly separating crystals from their mother liquor, as in the separation of sugar crystals in the preparation of granulated sugar. In this case the masse-cuite, consisting of syrup mixed with sugar crystals to form a paste, is discharged into a revolving

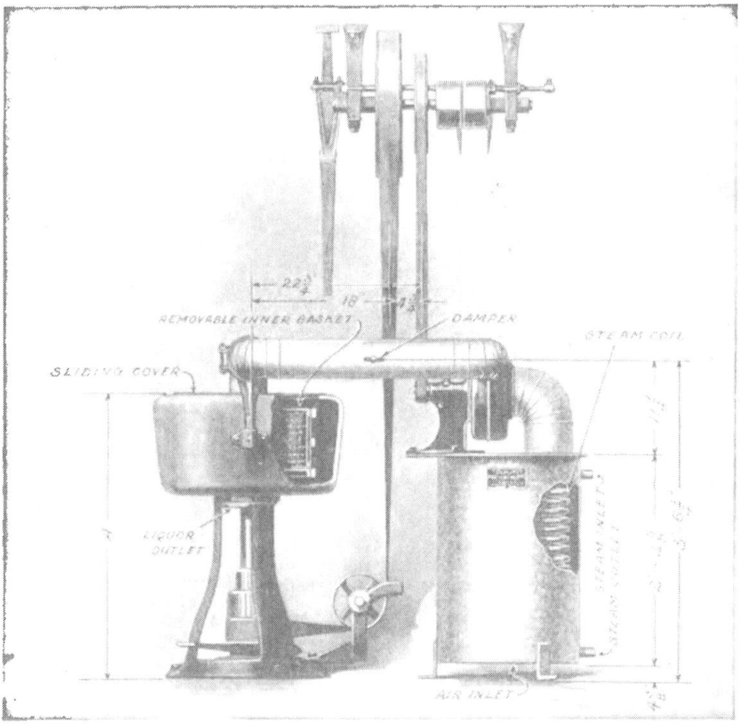

Fig. 107.

perforated drum. This may be under or over driven. Suspended over-driven centrifugals built by the Tolhurst Machine Works, Troy, N. Y., are shown in Figs. 105 and 106. In operating, the masse-cuite or other material is discharged into the machine which makes over 1500 r. p. m. for 10 minutes. A little water sugar solution is then sprayed on the sugar crystals. After allowing time to drain about 3 per cent. of water only will remain and the crystals may then pass to the dryer. The Tolhurst Co. also make an apparatus for drying small metal articles in the centrifugal shown in Fig. 107. This is used in plating plants, etc., and the centrifugal shown (Fig. 108) illustrates the under driven type power being supplied from overhead shafting through a belt.

(97) Fig. 109 shows an underdriven centrifugal constructed by Schaum and Uhlinger. In this cut 3 is the basket into which the material is placed. The machine is then started and the centrifugal force acting on the material causes it to climb up the sides

Fig. 108.

of the revolving basket, forming a wall on the inside. The crystals are prevented from passing over the top of the basket by the top ring 21.

The drum 27 is perforated and inside the drum there is a woven wire screen and inside the screen another finer woven screen, filter cloth or other filtering material according to the nature of the material to be dried. When the required degree of dryness is obtained, the machine is stopped and the valve 4, removed by means of the combined handle and locking device 15. The wall of material is then broken down and the crystals scraped through the opening 14 in the basket bottom 20, and flows out of the machine through the trough 12.

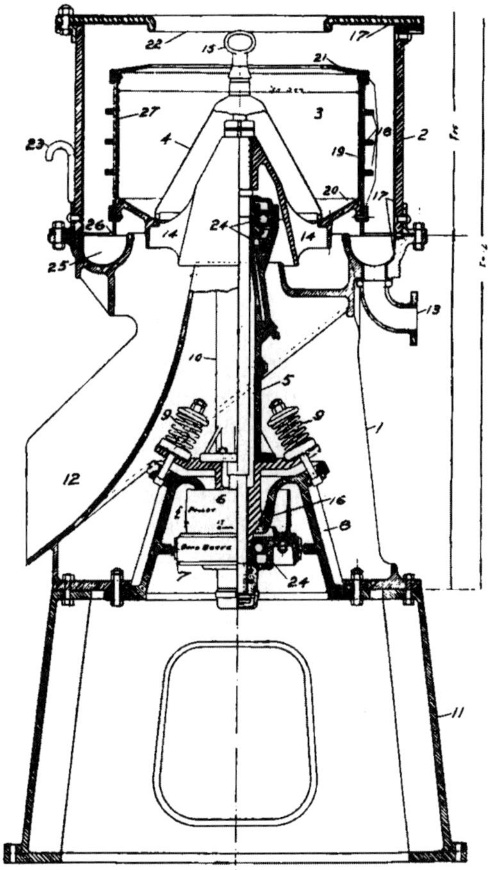

Fig. 109.

The liquid is thrown by centrifugal force against the inside of the curb 2, whence it runs into the gutter 25 and out through 13.

In commercial centrifugals the centrifugal force is 200 to 600 times greater than gravity.

"In practice the load can not be evenly distributed, or may not remain so due to the moisture content having been greater in one portion than another. As the centrifugal force is hundreds of times that of gravity it requires but a few pounds more on one side of the basket than on another to cause an unbalanced force of great magnitude. To avoid excessive strains on the bearing the basket is held flexibly by the ball 16 and the springs 9. With an uneven load the basket spins about its center of gravity and oscillates about the ball joint 16. In other words the basket runs out of true to an amount depending on how much the load is out of balance. The springs 9 do not affect the amount of oscillation to any appreciable extent, after the basket has acquired considerable speed, as the effect of the centrifugal force is too great to be overcome by the springs and their purpose is to relieve the bearings of excessive pressure. These springs are necessary in an underdriven machine to prevent the basket from falling over to one side. In an overdriven machine they are principally necessary to overcome the friction at the point of suspension.

"While the springs must be sufficiently strong to maintain the basket in an upright position, if excessively strong they introduce a disturbing element which is often encountered in high speed machinery, and is known as a "critical" speed. Under the influence of the springs, if the basket were thrust to one side and released, it would have a tendency to swing back and forth like an inverted pendulum until brought to rest by friction. The beat, or period of oscillation would depend on the stiffness of the springs on the one hand, and weight of the basket and its moment of inertia about the point of oscillation on the other.

"When the revolutions of the basket are such that they correspond with two beats of the basket as a pendulum, with an uneven load, we have the condition which produces a "critical" speed. Under these conditions the heavy side of the basket is revolved in synchronism with the swing of the basket about the spherical seat and the force which is stored in one spring when the heavy portion of the basket was on that side is given up just in time to be added to the effect of the heavy side of the basket on the spring directly opposite. The swaying of the basket would then increase until the conditions were changed or the swinging parts struck the stationary parts of the machine. As the speed is increased the revolutions and swinging of the basket no longer synchronize, and the basket begins to "spin."

"After passing the critical speed the heavy side moves in toward the center of revolution, and a piece of chalk held so as to just touch the outside of the basket would mark it on the most lightly loaded side. This phenomenon is rather difficult to understand, because the effect of centrifugal force is outward, and it would naturally be expected that the heavier side would swing the farthest from the axis of revolution. If it is considered, however, that the whole mass is being revolved without restraint, it will be seen that the inertia of the heavier portion will tend to cause the lighter portion to be revolved about it as a center. The result is that the center of gravity becomes the center of revolution, which explains the phenomenon.

"For chemical work, the parts of the centrifugal which come in contact with the chemicals must be of a material which will withstand chemical action. A variety of constructions are, therefore, necessary to meet different conditions. Baskets are made of iron, steel, bronze, monel metal, aluminum, etc. They are often coated with rubber, galvanized or tinned. The outer casings, likewise, are often lined with lead or rubber.

"When drying material in a centrifugal machine, centrifugal force is opposed to the capillary attraction of the solids for the liquids, and also to the adhesiveness of the liquids. The viscosity tends to make the drying slower. Owing to the capillary attraction becoming greater as the voids become smaller, we find the drying action to be slower and less complete with fine solids in a viscous liquid. With a liquid of low viscosity and solids of the degree of coarseness of ordinary granulated sugar and upward, the "purging" of the liquid is very quickly and completely performed."

(98) On account of the high speeds and the varying conditions under which centrifugals are operated, the greatest care should be taken to see that a machine is selected which is suitable for the particular purpose for which it is required. In a number of cases centrifugals intended for drying textiles, of which, owing to their bulky nature, but a comparatively small weight could be put in the basket, have been used for heavy chemicals, with the result of a serious accident.

Where subject to corrosion the machine, particularly the baskets, should be frequently examined to ascertain that they have not become weakened.

(99) Centrifugals are commonly used for drying cane, beet and milk sugar, common salt, ammonium sulphate, naphthalene, carbolic acid crystals, picric acid crystals, nitrocellulose, starch, soda crystals, di- and tri-nitrotoluol and for chemical crystals generally.

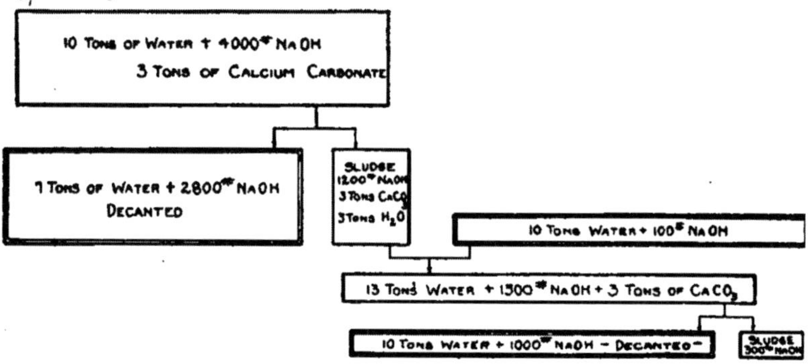

Fig. 110.

(100) Settlers.—The Dorr Co., of 101 Park Ave., New York City have applied an agitator and settler to the problem of separating solids from liquids. Their first work was in the treatment of gold ores with cyanide solution. This has been followed by other applications such as the manufacture of caustic soda from lime and sodium carbonate, the washing of sodium nitrate crystals, etc. A description of the caustic soda process will make this clear. The interaction involved is: $Ca(OH)_2 + Na_2CO_3 = 2NaOH + CaCO_3$. In order that this may reach completion from left to right it is necessary that the solution contain not to exceed 1 pound NaOH to 5 pounds water.

(101) "The areas of the rectangles shown in Fig. 110 represent the comparative dissolved content of the solutions used in the following description: The materials in the proper proportions are mixed with weak caustic solution (available from previous washing) in a tank so that after agitation the tank will contain three tons of precipitated calcium carbonate (dry) and 4000 pounds of caustic soda dissolved in 10 tons of water (400 pounds NaOH per ton of water). This mixture is allowed to settle until 7 tons of water (containing 2800 pounds of NaOH) together with the precipitate, remain in the tank in the form of a sludge.

"To this sludge is added a dilute solution available from washing a previous batch, consisting of 10 tons of water and 100 pounds of caustic soda (10 pounds of NaOH to a ton of water). The resulting mixture, after thorough mixing, contains the 3 tons of precipitate, 13 tons of water, and 1300 pounds of caustic soda (100 pounds per ton of water). Settling again takes place, and 10 tons of clear water (containing 1000 pounds NaOH) is decanted. There remains in the tank the 3 tons of precipitate and 3 tons of water containing 300 pounds of caustic soda.

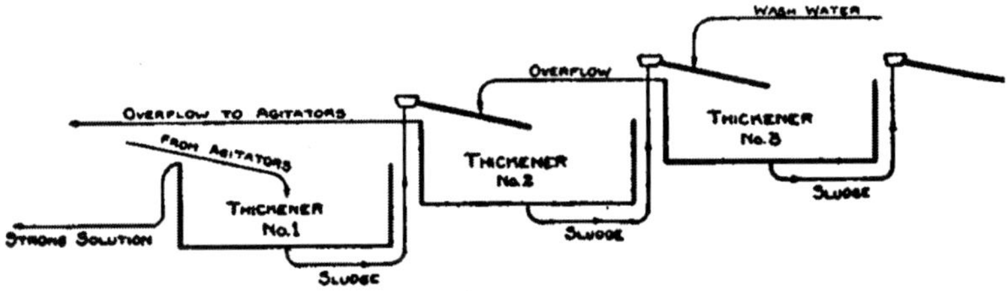

Fig. 111.

"By means of further steps along this line as above described, this sludge may be washed until practically all of the caustic is separated from the precipitate.

(102) "Considering this now from the standpoint of the continuous counter current flow sheet (Fig. 111), assume that the soda ash and lime mentioned in the preceding paragraph are agitated as above described, producing 3 tons of calcium carbonate (dry) and 10 tons of water containing 4000 pounds of caustic soda. Assume, too, that similar batches are produced at regular intervals of, say, two hours. At this rate there will be produced during 24 hours 36 tons of calcium carbonate and 120 tons of water containing 48,000 pounds of caustic soda, the whole of which mixture will be fed to thickener No. 1 at a uniform rate throughout the 24 hours. If the thickener were charged with a quantity of this mixture from previous batches, there would be overflowed, during each two hours operations, 7 tons of water containing 2800 pounds of caustic soda, while three tons of water containing 1200 pounds of caustic soda would leave the thickener mixed with calcium carbonate in the form of a sludge, exactly as was the case in the operation of the intermittently operated tank above mentioned.

"Now this 3 tons of water containing 1200 pounds of caustic soda, which leaves thickener No. 1 as part of the sludge during each two hours, is continuously transferred in like time to a second thickener, in which a wash is being continuously given the sludge by mixing it with 10 tons of water containing 100 pounds of caustic soda (the overflow from a third thickener). As in the case of the intermittently operated tank, the solution produced by this mixture consists of 3 tons of precipitate and 13 tons of water containing 1300 pounds of caustic soda. Of this mixture, 10 tons of water containing 1000 pounds of caustic soda will be decanted or overflow, going to the agitators to be used in making up a new batch, while 3 tons of water containing 300 pounds of caustic soda passes out in the overflow to be transferred to thickener No. 3 for a second wash.

(103) "From the above, the conclusion might be drawn that the same washing efficiency could be expected from the intermittent system as from the continuous counter-current decantation system, and theoretically this is true. In practice, however, this theory seldom holds good. In the manufacture of caustic soda, for instance, it is safe to say, that few of the operators using the intermittent system make an average recovery to exceed 96 per cent. or 97 per cent. of the caustic when using two washes. A plant using continuous counter-current decantation with Dorr equipment has shown a recovery of 99.5 per cent. to 99.6 per cent. over a period of several months' operation. This plant is producing a 15° Bé solution, and employs two washes.

(104) "In the continuous counter-current decantation system as generally installed to-day, the sludge coming from the thickener in the system is passed through a trough or launder, in which it is thoroughly mixed with the overflow from the subsequent thickener in the washing system. This mixing is one of the important features in the system and is responsible for the close approach to the theoretical washing efficiency made possible by it.

(105) "Fig. 112 represents a typical installation of Dorr thickeners and Dorr agitators in a system using continuous coun-

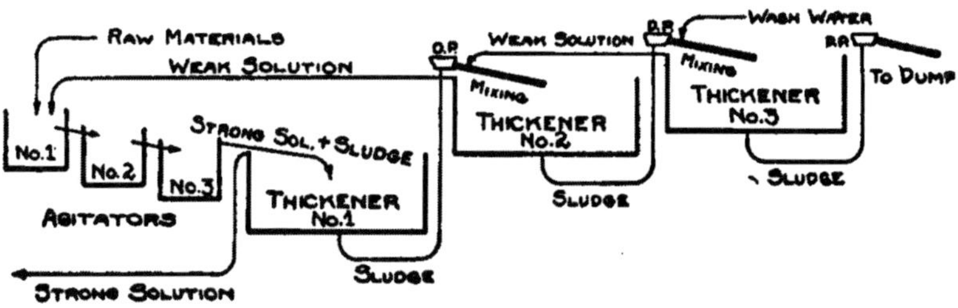

Fig. 112.

ter-current decantation and continuous agitation, suitable for the majority of chemical operations to which the system is applicable. The various machines are installed at the elevations shown so that the solution will flow from the overflow launder or trough of the one preceding. The overflow from thickener No. 2, which is the strongest wash water in the system, flows to the first of the series of Dorr agitators, where it is used as "make up" water for the reaction which takes place in the agitators. The mixture in which the reaction is taking place flows through each of the agitators in the series and finally to thickener No. 1. The overflow, or thickened sludge, from thickener No. 1 is transferred by means of a diaphragm pump, *DP*, to the feed launder of thickener No. 2 in which it is mixed with the overflow from thickener No. 3, as above described. The overflow from thickener No. 1, consists of the strongest solution in the system and is carried away for use or further treatment, as subsequent parts of the process may demand.

"It will be noted that the flow of the solution is accomplished entirely by gravity and that the only pumps required are used to control the rate of discharge from the various thickeners and at the same time transfer the thickened sludge from one thickener to the following one in the system.

(106) "One of the most vital points in the continuous counter-current system is the control of the moisture content and thereby, the amount of dissolved substance which leaves each thickener with the sludge. There is, of course, for each individual material a ratio of solution to solids below, which the material will not further separate by settlement, but it is essential that the underflow from each thickener have a consistency approximating

Fig. 113.

this ratio as nearly as is practicable. Several methods have been used to obtain this control of the underflow, but the use of the Dorrco pump (Fig. 113) has proven most satisfactory. The pump is connected with the thickener underflow and is located just above the side of the tank. This pump presents the advantage of a positive displacement through which an automatic regulation feed of discharge is approached, with a regular and definite feed to a thickener in a definite time. This pump is regulated by hand to give the displacement necessary for handling this volume. Since the volume of the solids is a definite quantity, the amount of solution which passes out with them in the above mentioned period is also very nearly definite in spite of any slight variations which may occur in pulp density. That such control is impossible with a valve or nozzle discharge is at once apparent. An air lift connected to the discharge pipe has been used in some installations, but a uniform density of discharge cannot be maintained with it. The body of the ordinary diaphragm pump is constructed of ordinary cast iron for use where such construction is permissible, but for acid solutions it is cast of Duriron or antimonial lead.

(107) "The Dorr Continuous Thickener consists of a slow moving mechanism made up of a central vertical shaft driven by a worm gear and worm, the shaft having radial arms attached to its lower end. These arms carry plow blades set at an angle which, through the rotation of the mechanism, move the thickened material to a discharge opening at the center of the tank. The mechanism is usually placed in a comparatively shallow, flat-bottomed tank into which the feed enters continuously at the center and from which the clear decanted solution overflows and is collected in a peripheral trough or launder. The overflow containing the solids is taken off at the center of the tank bottom and is usually controlled and elevated by a diaphragm pump.

(108) "The Dorr Thickener, Fig. 114, may be most simply termed a continuous settling tank, the action of the rakes or arms being so slow that in no way does it hinder settlement.

"The mechanism is so arranged that the shaft and arms may be raised by means of a lifting yoke attached to the top of the shaft, thus preventing the arms from being imbedded in the thickened material at the bottom of the tank in case the power should be shut off. The shaft may be gradually lowered while the machine is in operation, until the arms assume their normal position in the tank.

"The Dorr continuous thickener is usually equipped with an overload alarm which indicates the relative amount of power being taken off at any time, and, when the machine becomes overloaded, rings an electric bell. Immediate attention can then be given thereby preventing breakage.

(109) "The capacity of a thickener depends entirely upon the physical nature of the material handled and is governed largely by the area of the tank bottom. For each individual material there is a minimum area which must be available for each ton of dry solids to be continuously settled per 24 hours.

"The Dorr thickener may be built of acid resisting material where required.

Fig. 114.

(110) "**The Dorr Agitator,** Fig. 115, can be operated in a flat or conical-bottomed tank, and consists of a central vertical cylinder carried by a shaft supported from the top of the tank, and equipped with two arms carrying plows, as in the Door thickener, which travel around the bottom of the tanks and move the pulp to the center. The arms are hinged so that they may be raised to a sharp angle, thus obviating the trouble that might be experienced in starting up even in a sandy sludge, after which the arms can be lowered slowly to their normal position. The sludge or pulp is raised through the cylinder by means of compressed air and distributed uniformly over the surface of the mixture in the tank by means of suitable revolving launders.

Fig. 115.

"It is possible under certain conditions for the coarser particles in the tank to be given a longer period of agitation than the finer material by regulating the discharge point for the solids in the revolving launders."

Fig. 116.

Fig. 117.

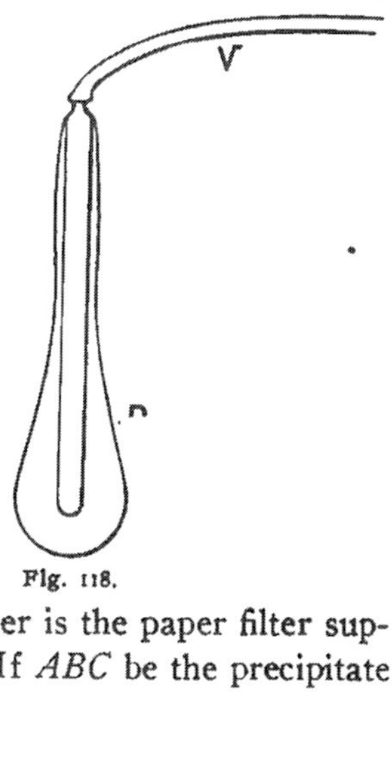

Fig. 118.

(111) Filters.—The most familiar filter is the paper filter supported in a funnel, shown in Fig. 116. If *ABC* be the precipitate left in the paper and we wash it by pouring water on the surface, the water will take the path of least resistance which is the shortest path, and travel from *A* to *B* and not from *A* to *C* so that no matter how often we wash it, the precipitate at *C* will be very poorly washed.

If we use the Shimer filter, shown in Fig. 117, consisting of a horizontal layer of paper felt supported by piano felt, and pour a precipitate upon it, we shall have a much better filter, capable of easy washing with a very small amount of wash water.

A third form of filter is shown in Fig. 118, which is a perforated cylinder covered with filter cloth, suction being applied at *V*. The liquid then passes through the cloth and the solid forms a cake on the outside. As all suspensions are densest at the bottom, even when stirred, the mass *P* is thickest at the bottom.

The mass is washed by lifting it into a vessel of clean water and again applying suction. The water again seeks the path of least resistance, and is more perfectly washed at the top than at the bottom.

(112) Filter Presses are constructed on two separate principles; they are run either by pressure or suction; an example of the first type which may be built either of wood or metal is shown in Fig. 119. This is one of the oldest types of filter press. It is rather slow working and requires considerable labor, but gives a dry cake containing from 10 to 25 per cent. moisture according to the nature of the material which may be pumped in until the press is full, by means of a pump or by forcing it through by compressed air. The last is the slower process, but necessary where the material acts on the pump. The pulp is first pumped in until the cells are properly filled and the cake is then washed by forcing water through it. The amount of wash water needed is relatively small, but depends upon the purity required in the insoluble material. Where very careful washing is required, it may be necessary to pulp the cake with water and filter and wash a second time.

(113) The press shown built by D. R. Sperry and Co., contains 50 recessed plates 42 inches in diameter, and has sufficient

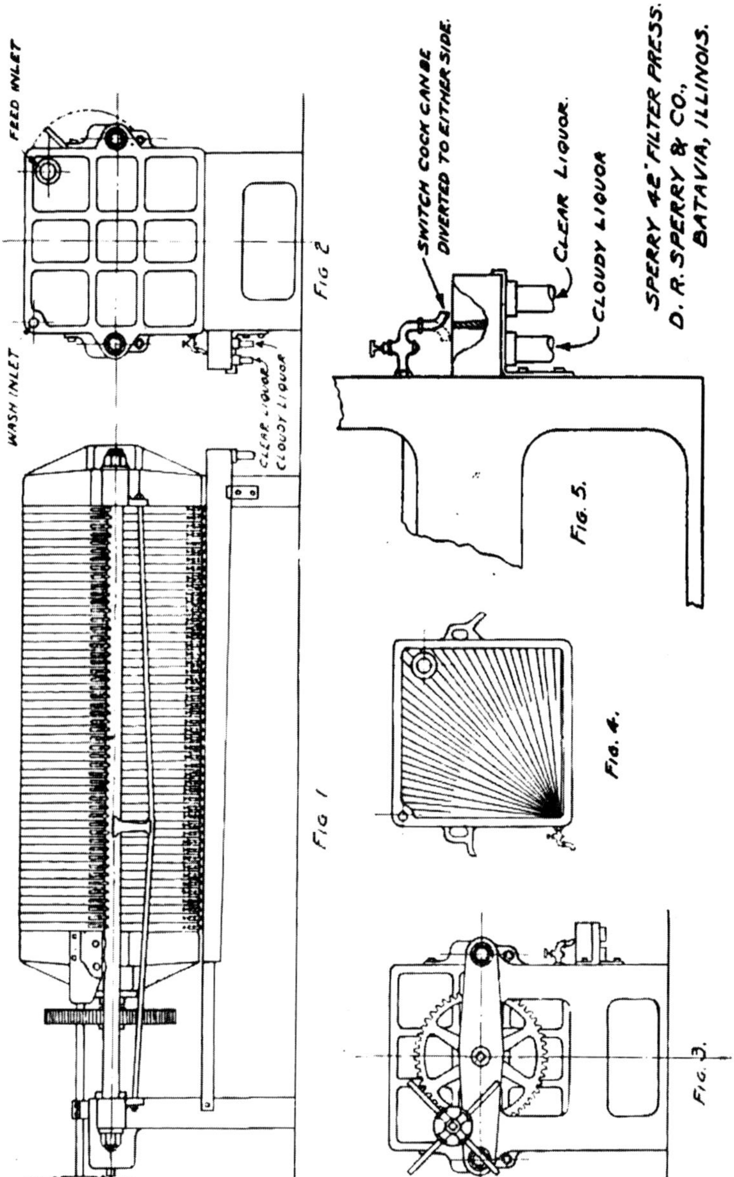
FEED INLET
WASH INLET
CLEAR LIQUOR
CLOUDY LIQUOR
FIG 1
FIG 2
FIG. 3.
FIG. 4.
FIG. 5.
SWITCH COCK CAN BE DIVERTED TO EITHER SIDE.
CLEAR LIQUOR.
CLOUDY LIQUOR
SPERRY 42" FILTER PRESS.
D. R. SPERRY & CO.,
BATAVIA, ILLINOIS.

Fig. 119.

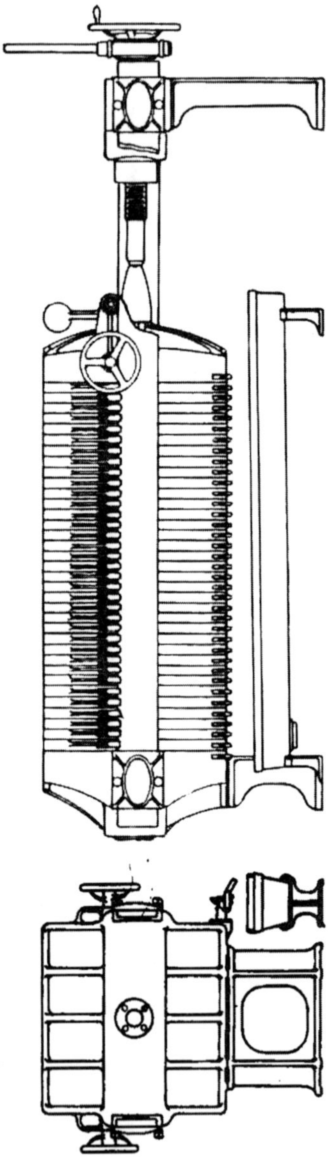

Fig. 120.

T. Shriver & Company,
Harrison, N.J.
36″ C.F.O.D. 36 Ch. Press.—Capacity 23.3 Cu. Ft.

capacity to filter 24 tons blanc fixe (precipitated $BaSO_4$) in 24 hours, or at the rate of one ton per hour, leaving the pulp with a moisture content of 30 per cent. which is required in the trade. This would take eight cycles, the press taking 3 tons at one operation. The labor of two men would be required for 1¼ hours during each cycle in opening and closing the press. This press costs at this writing (1918) about $2300 f. o. b. Detroit, Michigan. Each complete set of filter cloths of No. 10 duck will cost about $167 purchased in rolls and cut in suitable lengths by the press attendants. In running on blanc fixe the first runnings will be cloudy and must be run over again. Power required is that necessary to operate the steam pump 7x4½x6.

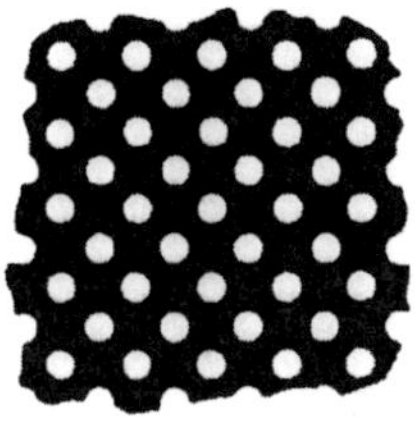

Fig. 121.

Fig. 122.

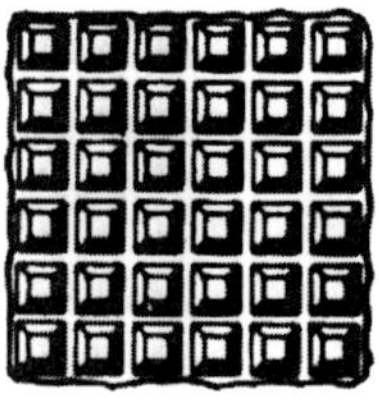

Fig. 123.

Fig. 124.

Fig. 125.

(114) Shriver Press.—Various forms of the Shriver Press are shown in Figs. 120-129. The perforated screen surface is shown in 121, corrugated surface in 122 and pyramid surface in 123. Shriver presses use the pyramid surface for which 70 per cent. efficiency is claimed. The flap cock used is shown in 124 closed and 125 open. The thrust block quick opening device is illustrated in 126, the ratchet closing device in 127 and 128. For the center feed press the tools used and the screwed clip nut cloth fastener are shown in 129.

Fig. 126.

Fig. 127.

(115) In the Kelly filter, Figs. 130-135, the leaves are contained in a horizontal cylinder into which the liquid is pumped

Fig. 128.

Fig. 129.

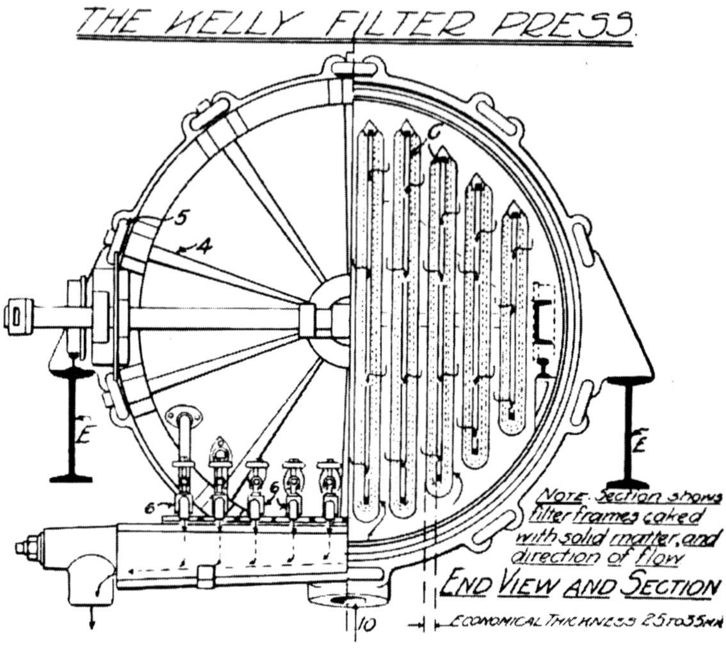

Fig. 130.

The Kelly Filter Press consists of a pressure cylinder "A" for containing the mixture to be filtered; a traveling filter-carriage "B" mounted on four wheels and carrying the filter frames; the filter frames "C" covered with bags of filter fabric; a quick locking device "D," mounted on the head of the pressure cylinder; and a trussed iron frame "E" supporting the entire filter press.

To operate the filter press, the carriage "B" is run into the pressure tank "A" and locked by means of lever "3" which forces locking-rods "4" into U-bolts "5." The mixture to be filtered is then admitted through slime valve "10" and rapidly fills the cylinder, entirely submerging the filter frames "C." Owing to the pressure at valve "10," the liquid is forced

THE KELLY FILTER PRESS.

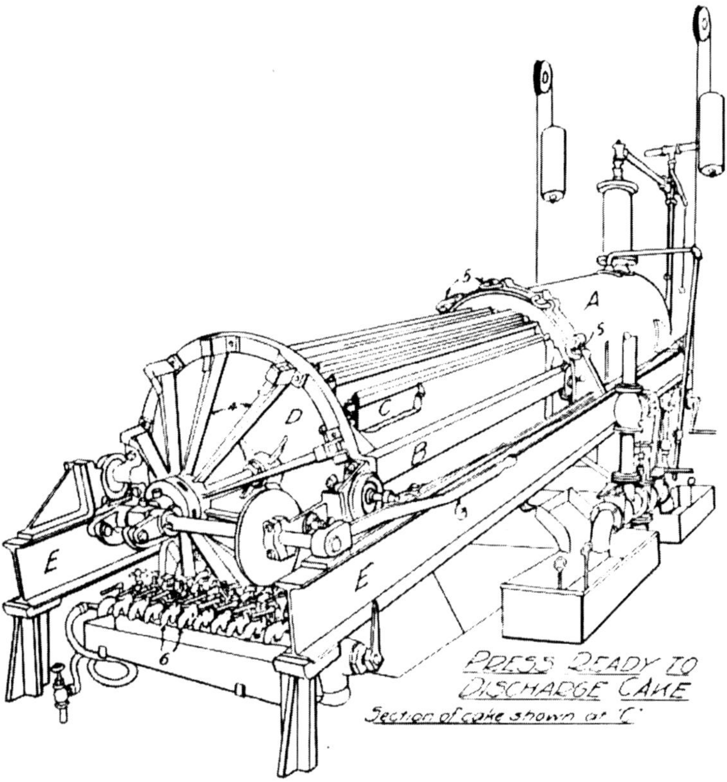

Fig. 131.

through the filter cloths, depositing thereon a cake of solid material, through which all remaining liquid must pass. The filtrate now continues through outlet valves "6" to the storage tank. As the thickness of cake *increases,* the rate of filtration *decreases* until an *economical cake thickness* is reached. At this point, filtration is stopped, the locking lever "3" is reversed and the added weight of the cake causes the filter-carriage to return down the inclined tracks to the position shown. The cakes may now be removed by steam, compressed air, or in any desired manner.

The advantages possessed by this press are: ease of operation, uniformity of cake, and clarity of filtrate.

for which purpose a montejus or centrifugal pump is preferred. When the cakes are of the proper thickness the liquid remaining

in the cylinder is discharged and the wash water introduced. After washing, the cakes are run out on a carriage and the mate-

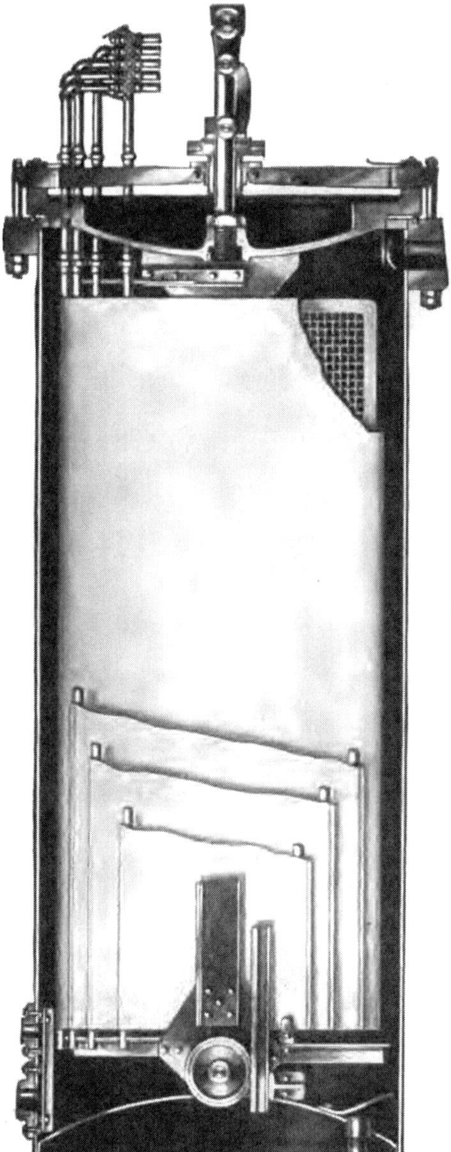

Fig. 133.

rial pushed off with a paddle. If a cloth is perforated, that leaf is closed by a cock provided for the purpose. The bag is then cut off and a new one slipped over the frame.

At present the Kelly filter is not used for acid liquids.

(116) **Sweetland Filter.**—In the Sweetland filter the leaves are first filled; the wash water then forces the excess sludge out and then washes the cake. This press requires more wash water and is more expensive to install, but because of ease of opening, dis-

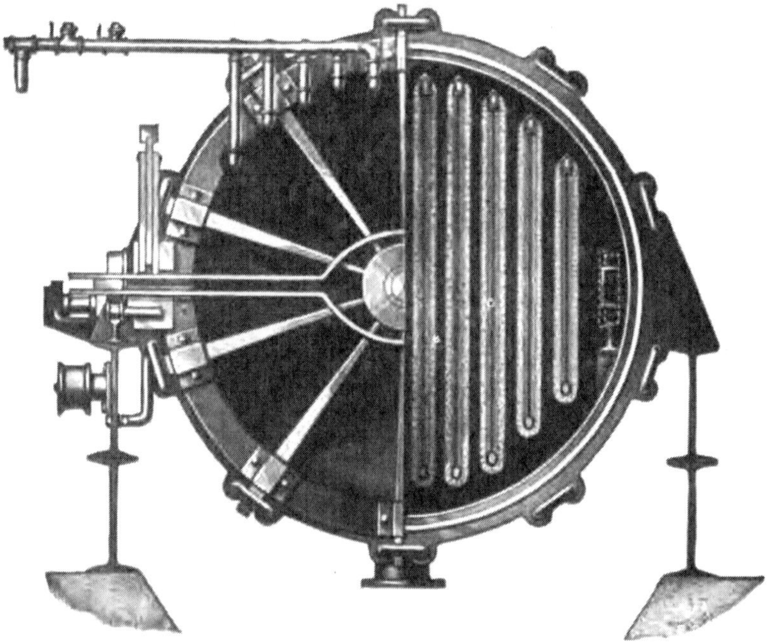

Fig. 134.

charging and closing, is more economical to operate. This filter is supplied with all parts lead lined or of bronze but not in wood. Steam jackets are supplied for hot filtration.

The body is cylindrical in form, divided on the horizontal center line into two halves. The two halves are hinged together. The hinge shaft has eccentric bearings to receive the hinges of the lower half. Since the bearings of the lower half are eccentric and those of the upper ones are not, the compression on the gasket at the back may be varied by simply turning the shaft. The front of the filter is closed and locked by means of a series of swing bolts mounted upon an eccentric shaft which rests in babbetted bearings along the front of the upper half of the filter body. This eccentric shaft tightens or loosens all swing bolts simultaneously by being turned through an arc of 90° at the same time the bolts are swinging outwardly. Figs. 136 and 137 show this filter open and 138 and 139 closed.

The main feed connection is at the center of one end (see Figs. 140 and 141), and a passage in the bottom casting distributes the feed uniformly. There are no flexible connections. An internal manifold pipe (see Fig. 141) with a series of nozzles opposite each leaf allows the operator to wash off any valueless precipitate and slide it out without opening the press.

The construction of the leaves is shown in Figs. 141 and 142. The sludge is forced into the filter at a pressure of 40 pounds, more or less, by means of a gravity pump or a montejus. The sludge rises in the press filling it completely and the valves which have been opened to allow air to escape are closed. The liquid now passes through the filter cloth and escapes through the outlet glass tubes, when the efficiency of each leaf may be continually watched, into the discharge pipe. The precipitate forms a nearly uniform coating on the cloth and then water is forced through until the cake is washed sufficiently. The cake is discharged by opening the filter and forcing air into the filters. When the precipitate settles rapidly it is better to use the

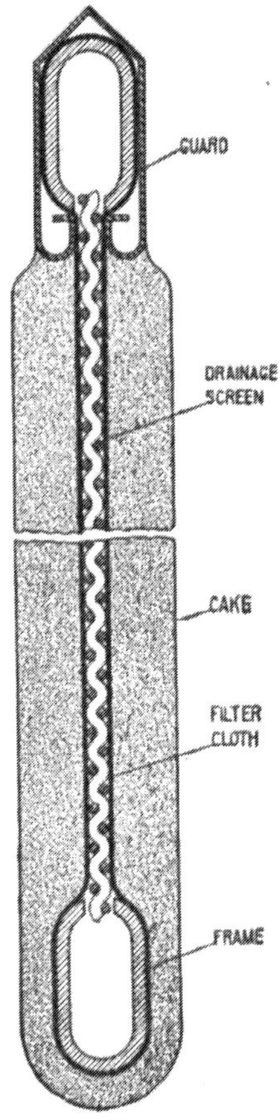

Fig. 135.

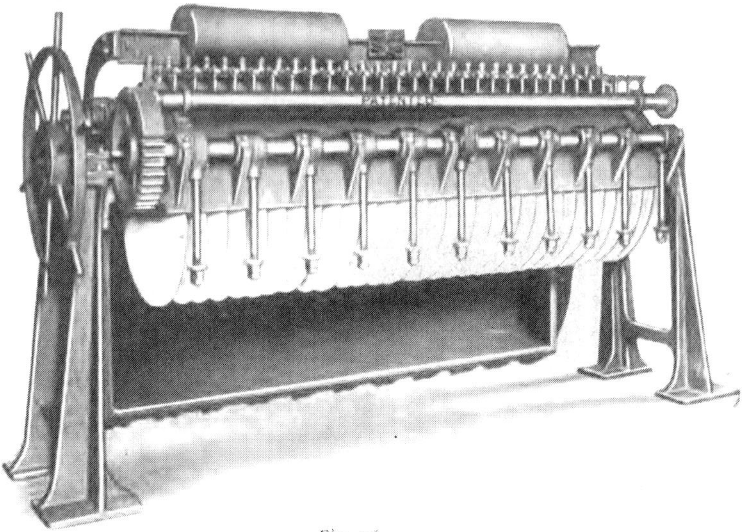

Fig. 136.

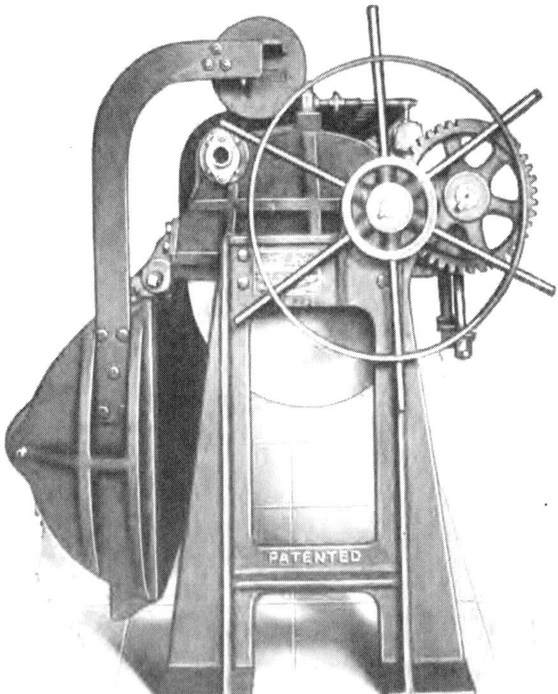

Fig. 137.

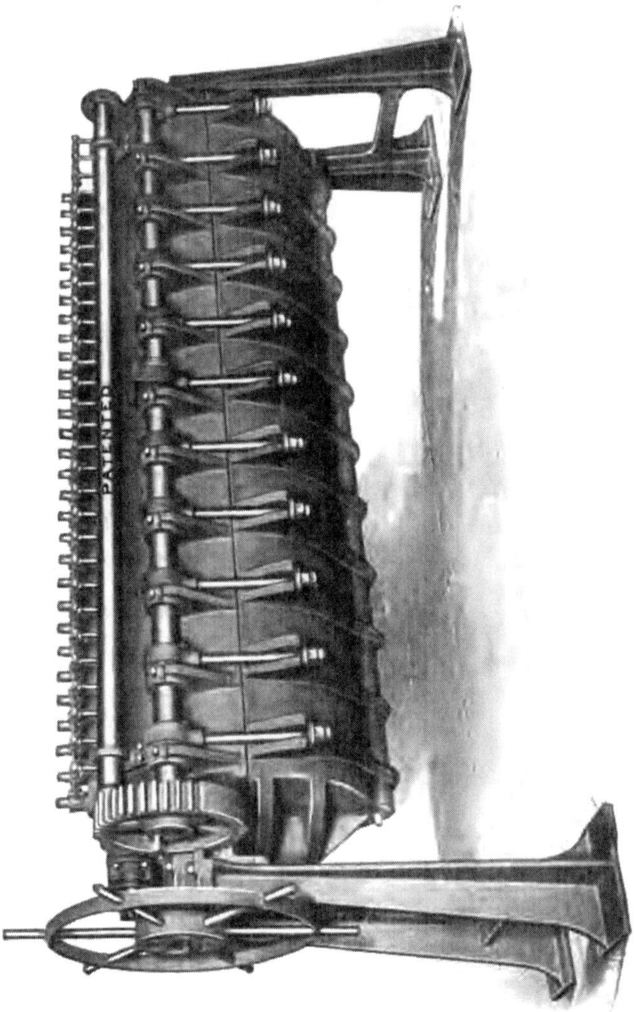

Fig. 138.

Sweetland rotatable filter shown in Fig. 143, in which the discs revolve constantly. A section of bronze trimmed lead lined filter is shown in Fig. 144.

(117) In the Moore filter the leaf is either flat or round and is dipped into the sludge. As suction is applied the liquid passes through the cloth while the precipitate forms an adhering cake.

Fig. 139.

This is washed by lifting it into a tank of clear water. Two waters are necessary as the first soon becomes fouled. Unless the sludge is stirred vigorously, it is densest in the bottom layers and the cake bridges over at the bottom layers of the tubes or plates. This causes incomplete washing and is one of the serious drawbacks of this type. The cake is also decidedly moist containing

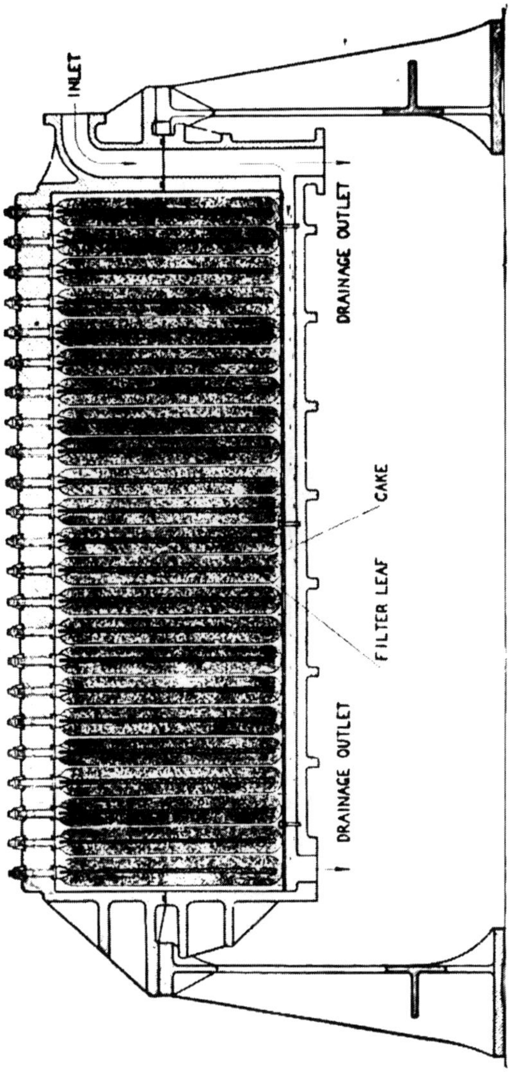

Fig. 140. Sweetland Filter. Longitudinal Section.

one-third to one-half its weight of water. In this, as in the Sweetland filter, discharging is effected by admitting compressed air behind the cake when it slides off.

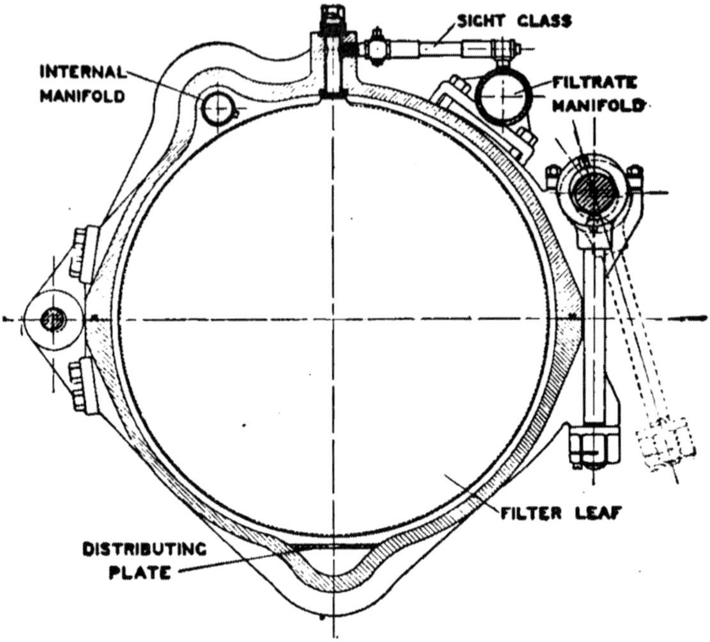

Fig. 141. Sweetland Filter. Cross Section.

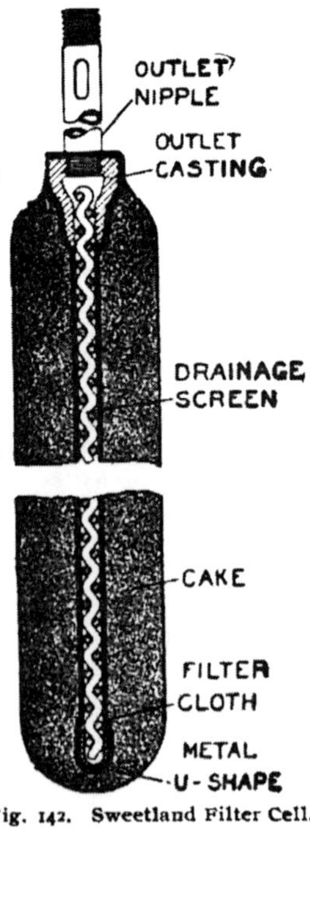

Fig. 142. Sweetland Filter Cell.

SWEETLAND ROTATABLE PRESSURE FILTER

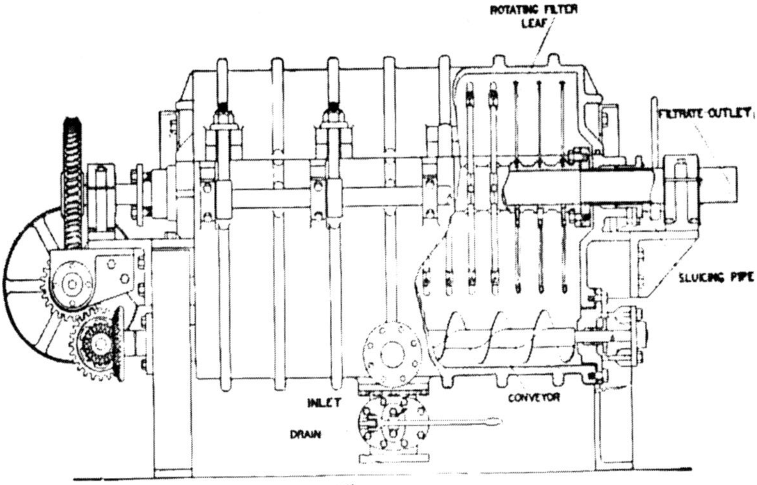

Fig. 143

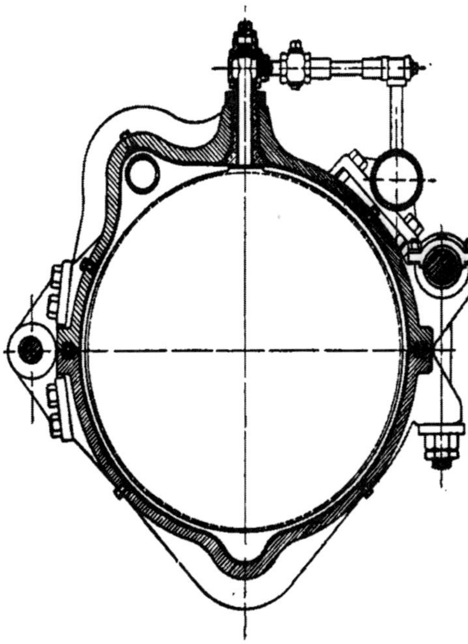

Fig. 144. Lead Lined Sweetland Filter.

AMERICAN CONTINUOUS SUCTION FILTER

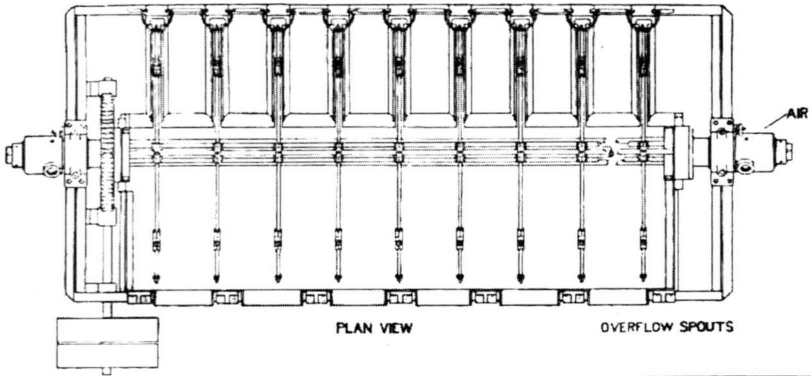

TYPE & SQ.FT	DIA OF LEAVES	NO OF LEAVES	LEAF SPACING	FLOOR SPACE	SHIPPING WEIGHT	A	B	C	D
	FT. IN.		IN.	FT.		FT. IN.	FT. IN.	FT. IN.	FT. IN.
65	4-0	3	11	6X6	3000	5-9¼	3-5	6-2⅜	4-11
130	4-0	6	11	9X6	4600	9-0⅞	6-11¼	6-2⅜	4-11
195	4-0	9	11	12X6	6300	12-3¾	9-8¾	6-2⅜	4-11
260	4-0	12	11	15X6	8000	15-0¾	12-5¾	6-2⅜	4-11
240	7-6	3	14	10X7	8300	6-7½	5-1¾	10-1⅜	8-7
480	7-6	6	14	10X12	11000	10-1½	8-8½	10-1⅜	8-7
720	7-6	9	14	15X10	14000	14-8	11-8½	10-1⅜	8-7
980	7-6	13	14	20X10	17000	20-0	17-0½	10-1⅜	8-7

BOTH TYPES 9 & 12 LEAF FILTERS ARE PROVIDED WITH TWO SUCTION VALVES

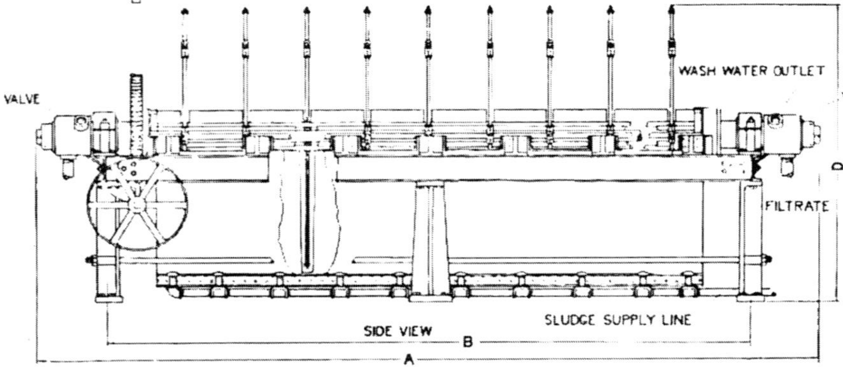

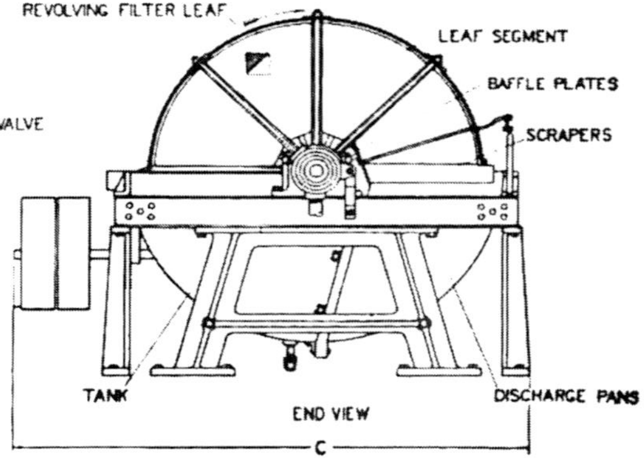

Fig. 146. American Filter.

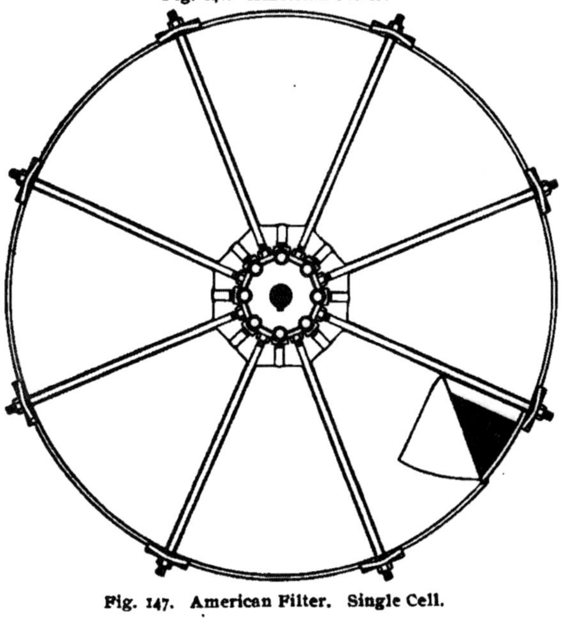

Fig. 147. American Filter. Single Cell.

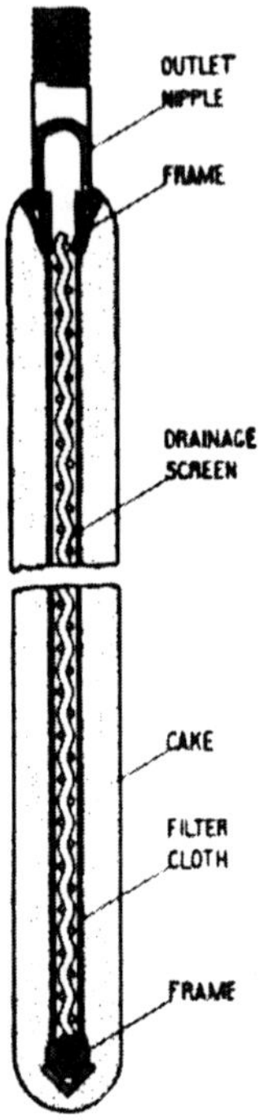

Fig. 148.

(118) In the American Continuous Suction filter, Figs. 145-149, which like the Kelly and the Sweetland filter, is manufactured by the United Filters Corporation, each disc or wheel is made of eight segments clamped to the shafts by rods. The shaft has eight longitudinal shaft compartments (Figs. 145-147). At one end of the shaft is the driving mechanism for rotating the discs, at the other the valve (Fig. 149). The shaft revolves clockwise viewed from the valve end. The valve is conically ground and has eight ports corresponding to and connected with the longitudinal shaft compartments. Port No. 1 on the underside connects with the four submerged leaf sections. No. 2 connects with the wash water suction and dumping line accommodating the three upper left hand leaf segments. No. 3 admits compressed air into each segment as it passes the port. The precipitate in the sludge deposits as the segment moves through the pan. The filtrate discharges through the same opening from four segments in the sludge. As any given section lifts, the filtrate port closes and a wash water port opens and wash water is sprayed over the surface of the cake in very fine spray, the wash water having a pressure of 60-70 pounds.

The spray nozzles are adjustable and are placed near the periphrey of the moving discs. The washing is continued through 45 degrees. The moisture is then sucked out and the cake discharged by air pressure and deflectors shown in Fig. 141.

(119) In the Oliver filter shown in Fig. 150 the material is likewise formed into a cake by suction. The cake is washed by a water spray and then scraped off as the drum revolves. This apparatus gives a drier cake than the Moore filter and since it is continuous in operation has large capacity. It is expensive to install.

(120) The Portland filter, Fig. 151, resembles the Oliver filter.

(121) Filter frames are of two types, the flush and recessed. The cake produced varies in thickness from 1 to 6 inches according to the material, being thicker as it is more permeable.

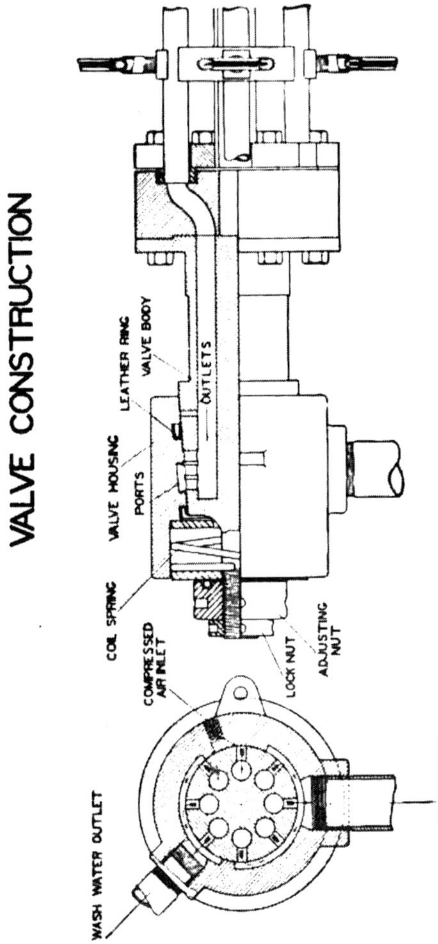

Fig. 149. American Filter.

(122) In the recessed type of filter the material is pumped in through an opening sometimes in the center, sometimes in a corner of the plate which may be either round or square. The cloth is held in place by a lock nut screwed down over the open-

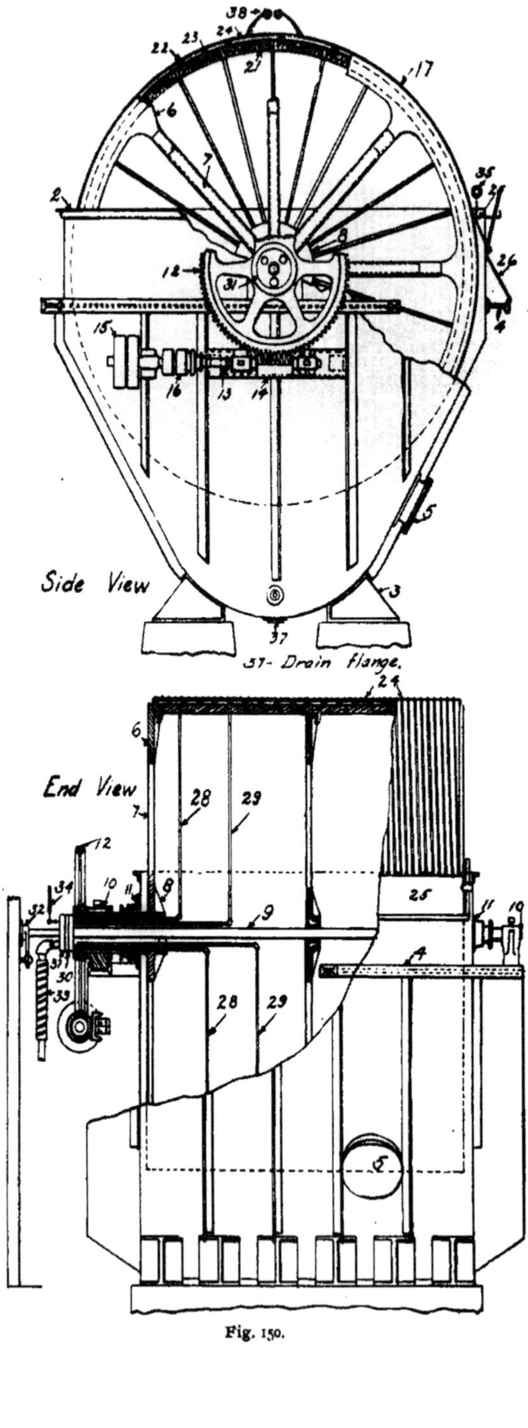

Fig. 150.

The Oliver Continuous Filter.

The vaccuous space 1, is covered with canvas protected by a wire winding on the outside. This space is connected with the air-pump by means of the pipes 28 and 29. The drum is slowly turned, in the tank 2, containing the pulp to be filtered, by means of the worm 14. The cake is washed by a water spray at 38, and is scraped off at 25.

1. Vaccuous space.
2. Steel filter tank.
3. Cast-iron pedestals.
4. Steel I-beam frame.
5. Manhole.
6. Cast-irou spider rim.
7. Channel steel arms.
8. Hollow trunnion.
9. Steel shaft.
10. Main bearings.
11. Stuffing boxes.
12. Worm drive gear.
13. Worm shaft.
14. Oilwell for worm.
15. Filter drive pulleys.
16. Highspeed wiring pulleys.
17. Filter drum.
21. Wood stoves for drum.
22. Section division strips.
23. Filter medium.
24. Wire winding.
25. Steel scraper.
26. Tailing apron.
28. Vacuum pipes.
29. Compressed air pipes.
30. Regrinding valve seat.
31. Automatic valve.
32. Adjusting lever for valve.
33. Vacuum connection.
34. Comp. air connection.
35. Discharge spray pipe.
38. Wash-water pipes.

ing. The pulp is forced in through the head of the press and the liquid passes through the filter cloth upon which the solid deposits

Fig. 151. Portland Filter.

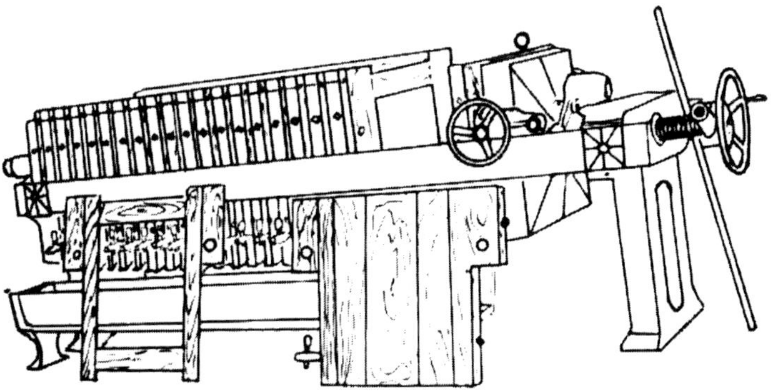

Fig. 152. SIDE FEED WOODEN CHAMBER FILTER PRESS.

and out through a cored opening into a side trough until a proper thickness has been attained. In some cases the recess is filled

LATEST IMPROVED VARNISH FILTER PRESS.

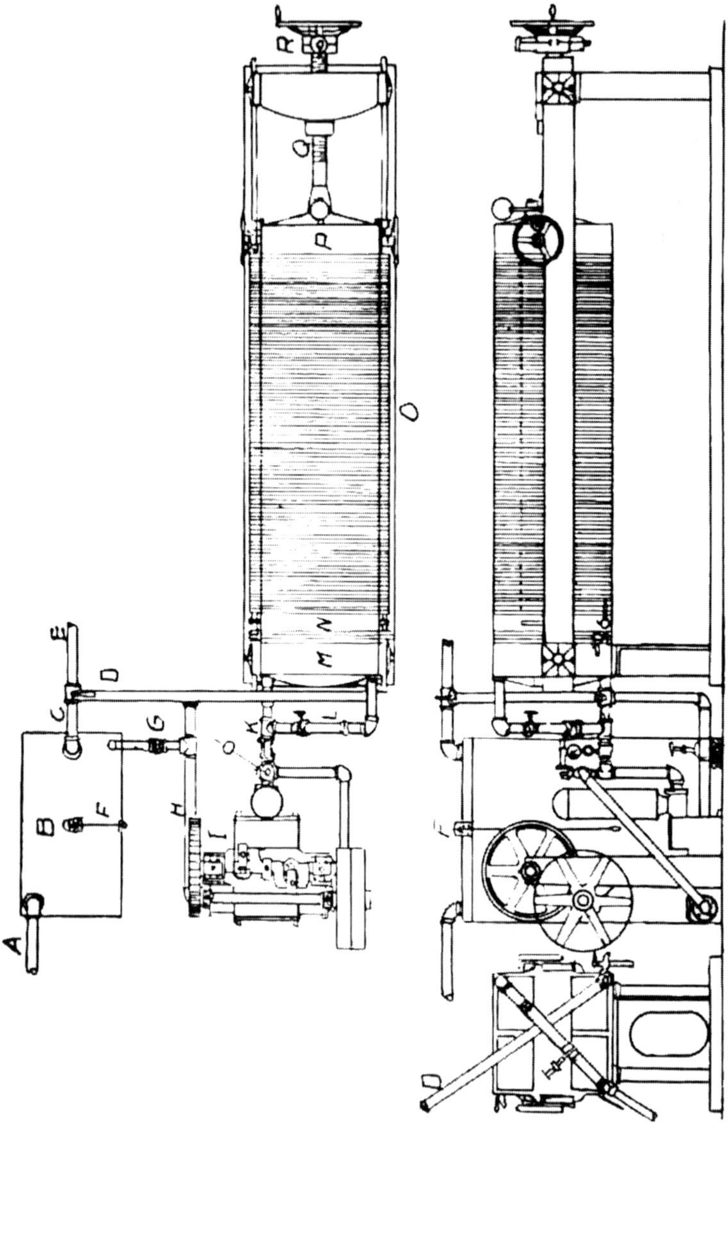

Fig. 153.—Varnish Filter Press.

full of material and the cake is not washed, as in compressed yeast manufacture. In other cases the supply of pulp is discontinued and is followed by wash water which displaces the mother liquor. The course to be pursued depends upon the material operated on.

(123) In the flush type the plates having a filtering surface alternate with the open frames filled from a cored lug at one side or at one or two corners. The discharge is through cored openings in the closed frame. The filtering surface may consist of a perforated screen or a corrugated or pyramid surface. Where acid liquids are to be filtered a press with wooden frames of the flush type shown in Fig. 152 must be freed. Fig. 153 shows a press used for filtering varnish.

Where wax or tallow is to be filtered cored openings for the circulation of hot water or steam through the frames are provided and where the liquid must be kept cold, as in filtering castor oil, cold brine is the circulating medium.

The capacity of filter presses is small compared with their cost and the labor charge high. They are commonly regarded as a dirty, sloppy necessity.

(124) Prominent makers of filter presses of the types enumerated are:

T. Shriver & Co., Harrison, N. J.
D. R. Sperry & Co., Batavia, Ill.
United Filters Corp., New York City.
John Johnson Co., 39th St. & 2nd Ave., Brooklyn, N. Y.
The Albright-Nell Co., Chicago, Ill.
Roberts S. Refield, 327 S. LaSalle St., Chicago, Ill.
Wm. R. Perrin & Co., Old Colony Bldg., Chicago, Ill.
Oliver Filter Co., Market St., San Francisco, Cal.
Industrial Filtration Corp., 115 Broadway, New York City.
Colorado Iron Works Co. (The Portland Filter), New York City.

(125) The amount of water remaining in the press cake varies from 10 to 50 per cent. according to the apparatus used, the pressure and the nature of the material. The closed pressure usually gives the best results in this respect.[1] In a suction filter not more than 16 pounds at the best can possibly be obtained. In the positive pressure this is often increased to 120 pounds or even more.

(126) A material called Filtros, consisting of silica grains of various sizes cemented at 2600° F. with a fusible silicate has lately been put upon the market as a filtering material by the General Filtration Co., of 326 Cutler Bldg., Rochester, N. Y. This is supplied in plates and tubes of various sizes and grades. The proportion of voids in the material is claimed to be from 33 to 40 per cent. It is insoluble in acids and unaffected by chemicals which do not act on silica. It is manufactured with varying porosity to suit different conditions and has sufficient strength, a plate 12x12x1½ inches supporting a weight of 400 pounds without fracture.

(127) The same company manufactures a filtros wheel composed of filtros plates. This is divided into eight segments each having two compartments. Each compartment is completely isolated from the others. This is operated as a vacuum filter while the compartments are immersed. One revolution in 3 minutes is about the rate best suited for a lime sludge with 36 per cent. solids and in this time about a ¾-inch cake will be formed. As the cake emerges from the sludge it may be washed and is then discharged by about 3 pounds air pressure upon an apron above the tank. About 4.5 tons per hour of dry solids are produced, the cake containing about 40 per cent. moisture. The Kieselguhr Co. of America, manufacture a similar material called Filter-Cel.

[1] A recent paper by E. Hatschek is noted (*J. Soc. Chem. Ind.*, **39**, 2267) as a good illustration of the dogmatic method. Some, at least, of the statements made appear to be incorrect. (See also *Chemical Age*, **3**, 67, 1920.)

CHAPTER IX.

TANKS.

Containing vessels in the chemical industry are commonly spoken of as tanks and may be made of various materials, the cheapest being made of wood.

(128) Wooden Tanks.—Large round wooden tanks should always be placed vertically so that the bottom pressure is supported by the foundation. With a tank of, say 12 feet diameter, placed horizontally and filled with liquid, the pressure will force the ends out of the croze and finally destroy the tank. The ends may in such cases be supported by braces across the ends tied together by rods through the tank, but this is expensive and not very satisfactory. Tanks are best made round as the heavy pressure can be perfectly supported by hoops as shown in Fig. 154.

(129) Cypress gives no taste to cider vinegar or fruit syrups and is best for hot acids and liquids. Douglas fir is used in vinegar plants. Poplar is best for whiskey, as it is close grained and lasts well. It does not last well with other liquids. Yellow pine containing as much resin as possible is best for strong cold acids and chemicals. White pine is best for water, brine and pickle.

It is well known that strong acids and some salt solutions act on wood. Weak acids are less vigorous and attack first the lignin so that the cellulose fibers are set free and the wood disintegrates. At high temperature this action is more rapid. (See also *Met. & Chem. Eng.*, **18**, 528.)

Sizes of Tanks and Parts Thereof.

A. Thickness in rough	1½″	2 ″	2½″	3 ″
B. Finished thickness of staves	1⅜	1¾	2¼	2¾
C. Depth of croze	⅜	½	⅝	⅝
D. Width of croze	1⅛	1⅜	2	2½
E. Length of chime	3⅜	3⅜	3½	3½
G. Finished thickness of bottom	1⅜	1⅞	2⅜	2⅞
H. Thickness of beveled edge	1 5/32	1 21/32	2 1/32	2 17/32
J. Thickness of bevel	7/32	7/32	11/32	11/32
K. Length of bevel	1⅜	1⅜	1⅜	1⅜

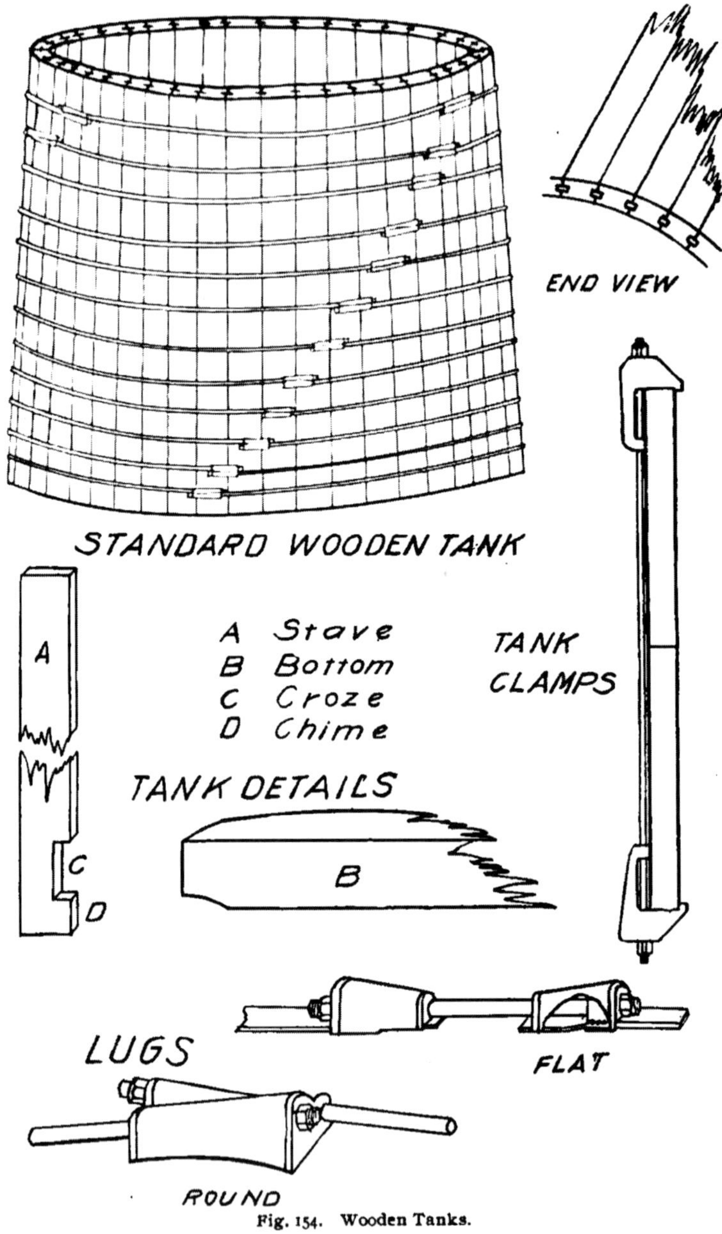

Fig. 154. Wooden Tanks.

No.	Capacity in Gallons Inside diam. ft.	Taper 1 inch to 1 foot Inside depth ft.	No. of hoops
1	3	2½	3
8	4	3	3
16	5	4	4
23	6	4	4
40	7	8½	7
55	8	8½	7
67	9	8½	7

Large tanks are shipped knocked down and must be set up where they are to be used. They should have concrete or post and wooden sill foundations and since there is sure to be some leakage, this should be collected in a sump. One cement floor with one sump provided with a proper pump answers for a number of tanks. Where acid liquids are in use, the collecting floor should be covered with 1½ inches of Johns-Mansville acid proof mastic or some similar material.

(130) Hoops for Tanks.—Hoops for wooden tanks are almost always made of round steel. This form has the great advantage that the side next the tank can be inspected and painted when needed. Flat hoops may rust very badly on the inside before this is detected and must be taken off entirely before being painted. If the tensile strength of the steel used is 60,000 pounds per square inch, the allowable strain is 15,000 pounds. The allowable strain for

½	inch	hoops	1,650	pounds
⅝	"	"	3,315	"
¾	"	"	4,230	"
⅞	"	"	6,090	"
1	"	"	8,355	"
1⅛	"	"	10,350	"

To show how the hoops should be put on let us suppose we have a tank 12 feet in diameter and 24 feet high with hoops of ¾-inch steel with tensile strength and factor of safety as above.

(131) The pressure exerted by water is due to its weight or 62.425 pounds per cubic foot. On the bottom of a 24-foot tank the pressure would be $62.425 \times 24 = 1498.2$ or for a 12-foot tank $1498.2 \times 12 = 17976$ pounds, or for each inch of depth $\frac{17976}{12} = 1498.2$ pounds. This strain is divided between the two sides of the hoop and would be 749.1 pounds on each rod. The ¾-inch hoops must, therefore, be spaced $\frac{4230}{7491.} =$

5.6467	inches at	24	feet
5.8923	"	23	"
6.1601	"	22	"
6.4534	"	21	"
6.7761	"	20	"
7.1327	"	19	"
7.5290	"	18	"
7.9719	"	17	"
8.4702	"	16	"

When the liquid is heavier or lighter than water, these distances will be altered in proportion to the change in specific gravity.

(132) Wooden tanks for muriatic acid may be made of 2-inch yellow pine of any convenient length but not over 6 feet diameter, placed horizontally. They are slip tongued and grooved and the grooves treated with slaters cement. The hoops are of steel and must be extra strong. The ends are recessed into double plank covers with slater cement between the plank and in the grooves. The heads are bolted on with rods passing the length of the tank outside. They must be well painted. The inside is painted with two coats of asphalt paint.

These tanks may also be made by first boiling tar until all the water and part of the volatile products are removed and then the staves (best of redwood) are immersed and boiled for about an hour. The staves are then used to make up the tanks. In boiling the tar great care must be used or a troublesome fire may occur.

A. Schwarz (No. 746,748, Dec., 1903) has patented the use of two tanks, one set within the other with the intervening spaces filled with pitch or other suitable material.

(133) Lugs for tank hoops are made of malleable iron. List prices range from 30¢ for ½ inch to 80¢ for 1-inch rods. They are shown in Fig. 154.

FOUNDATIONS.

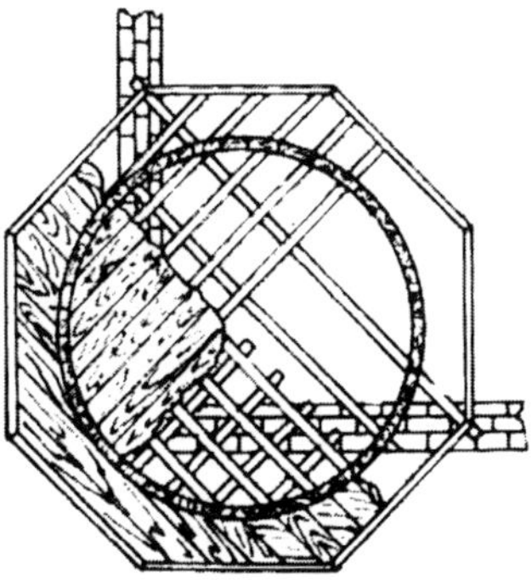

On Corner Walls.

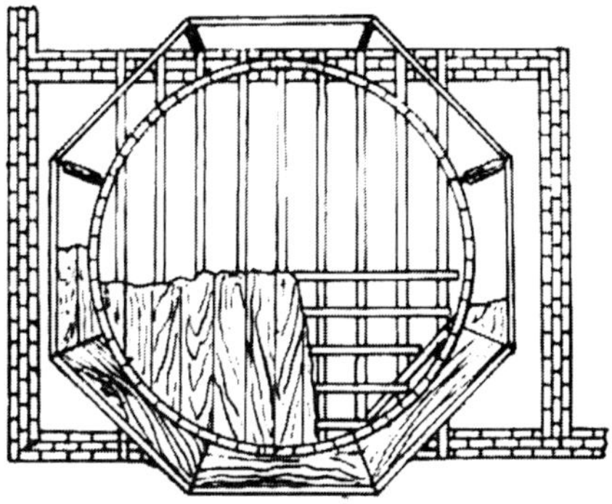

On Stair Tower.

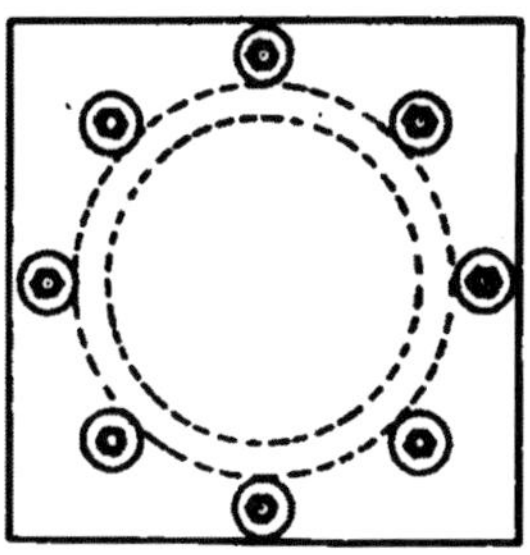

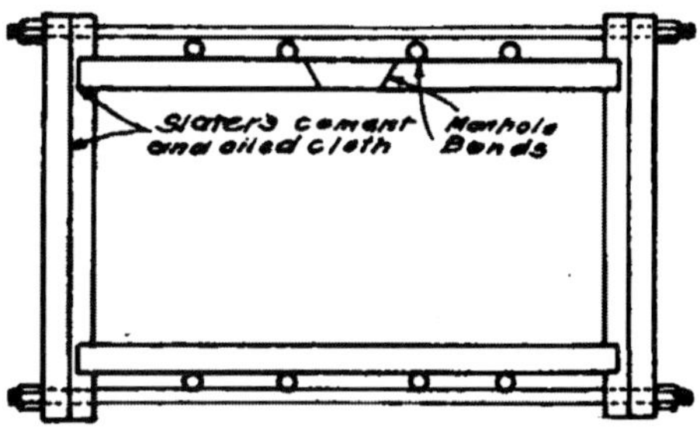

Storage Tank for Acids.

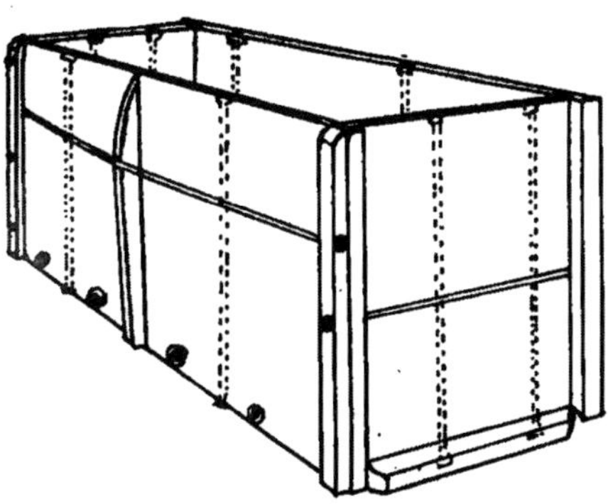

RECTANGULAR TANK.

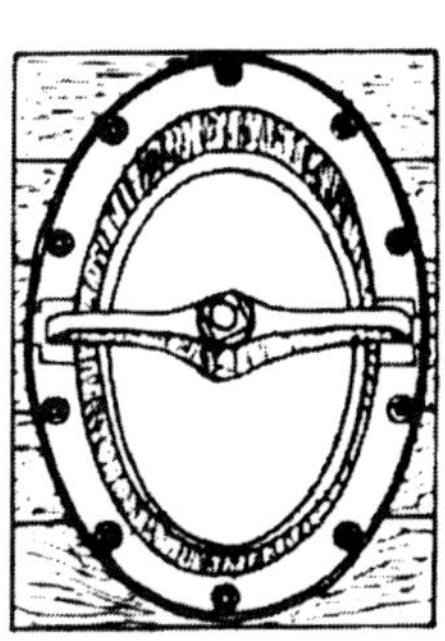

Patent Manhole Door.

Fig. 155. Wooden Tanks. Detail.

(134) **Rectangular Tanks or Vats** are troublesome since they cannot be kept as tight as round ones. When it is necessary to use them they must be drawn up in all directions by means of bolts run through the wood as shown in Fig. 155. A cheaper clamp, fastening on the outside of the wood is also shown in Fig. 154.

(135) In 1902 I saw a large number of cheap tanks at Daggett, Cal., used for the evaporation of boracic acid solutions by means of solar heat. Since these tanks may be useful occasionally in other localities, I have thought it well to describe them: A level piece of ground was chosen which was cleaned of stone and rubbish, covered with a few inches of sand carefully leveled. This was divided up into squares or rectangles of the size of tank wanted by stout stakes driven into the soil. Upright rough boards 1 foot wide were then nailed inside the stakes. A layer of tar paper with the edges overlapping was then spread on the floor and the ends and edges carried up over the sides and tacked fast. A layer of hot pitch was then painted over the tar paper and another layer of tar paper laid over it so as to break joints. When three or four layers of tar paper had been put on in this way the tank was pronounced finished and the liquid run in. Such tanks can probably in some cases be used for crystallizing, but must, of course, be carefully tested in each case before laying out expensive installations. In some cases it is necessary to shovel out the crystalline contents of such tanks. To protect the bottom, gratings of wood, perforated metal or coarse wire cloth may cover and protect the floor. Concrete tanks usually require no such protection.

(136) **Concrete Tanks** are more desirable for crystallizing, where it is possible to use them, since they lose heat more rapidly than wood. They are best made with round reinforcing, inside a rough wooden frame with steel hoops on a concrete floor provided with a sump and pump. Inside this another frame of ½-inch stuff is placed. About six inches on one side is left open, having wooden reinforcements on each side. Across this opening stretch turnbuckles. This frame is so diminished that a 3- to 6-

inch opening is left between the inside and outside frame, which is filled with *carefully rammed* concrete made of sharp sand and cement. The opening on the inner frame is closed with a removable board. When the cement has set, say in three or four days, the board is taken out; by turning the turnbuckles the inner frame is drawn together and may be removed and again used. The outer framework is also removed.

(137) Rectangular concrete tanks may also be built, but are not very strong unless braced on the outside. Where running water is available, crystallization may be hastened by using pipe as the reinforcement in such tanks, placed near the inside, and running water through these pipes. Crystallizing tanks of whatever material are best made with hopper shaped bottoms provided with a drain pipe and drop door so that the crystals may first be drained and then shoved through the open door into a centrifugal. This decreases handling materially. Browne[1] calls attention to the fact that concrete tanks should be built of slow setting cement, which must stand the pat test. It is best mixed with crushed granite. A tank 3 feet wide, 9 feet long and 2 feet deep costs about $12. It may be built on a wooden floor on tar paper and should have a 6-inch bottom and 4-inch sides. The inner frame of such a tank is made by putting square pieces in the corners and fastening the side boards with angle irons. When these are unscrewed, the boards drop in and are removed. Caustic or chloride does not act on cement nor does chlorine. It may be coated with hot paraffine, tar or asphalt varnish or mastic to prevent action by acid liquids.

(138) A good method for constructing cement tanks is to build a framework of expanded metal pipe or rod and then cover this with cement as plastering is done. This makes a cheap and efficient tank. Such tanks may be used, for example, as containers for nitric or sulphuric or mixed acids. Dr. Arthur Lackman, (*Trans. Amer. Inst. Chem. Eng.*, **8**, 110) describes a method of making cement tanks only 2 inches thick which are, however, very expensive. A form is first made of iron bars, spaced approxi-

[1] *Electrochemical Industry*, Vol. I, p. 273, Apr., 1903.

mately 1 foot apart and covered with wire cloth. The paste made of three sand and one cement is then plastered on in layers about ½ inch thick, first inside then outside until proper thickness is used. The tanks were 10 feet by 10 feet and were used for fermenting wine. Wine has a tendency to make cement waterproof. Seepage may be corrected by soaking with lime and water and allowing carbonation to take place. Attention is also called to two papers on the building of cement tanks using the apparatus controlled by the Cement Gun Co., of Allentown, Pa., which appeared in the *Engineering News* of November 9, and the *Engineering Record* of October 21, 1916. In this way the tanks of large diameter subject to considerable heat may be built with very thin walls. It is stated that a new material especially adapted for this purpose has recently been put on the market by the Trussed Concrete Steel Co. It consists of rods netted together with wire on the order of expanded metal with larger longitudinal members which have the effect of hoops.

(139) Albert Moyer[1] recommends painting such tanks with a dilute sodium silicate solution applied with a brush. After drying, wash with water, dry and then apply a second and then a third coat. Then wash and dry.

(140) Anson Betts[2] (No. 918,649, Apr. 20, 1909) has patented the filling of cement tanks with melted sulphur by submerging the previously dried tank in a pot of melted sulphur which is heated to any desired temperature.

(141) Wooden, cement or steel tanks lined with brick laid in melted sulphur or in Duro cement are often used. The latter cement seems to consist of silicate of soda and silica. It is washed with 60° Bé. sulphuric acid to set it properly, applied with a spray or rubber brush. Duro brick and the Duro cement are supplied by the Harbison-Walker Co., of Pittsburgh.

(142) A Pachuca or Precipitating Tank is shown in Fig. 99. It is made of wood the bottom being filled with cement to form an inverted cone, which may be lined with wood if necessary.

[1] *Met. & Chem. Eng.*, 8, 702, Dec., 1910.

[2] *Ibid.*, 7, 248, June, 1909.

The air jet is used for mixing the precipitating liquids. One or both the materials may be added in the dry state. For example, if dry lead acetate and sodium sulphate be mixed with the proper amount of water in the pachuca and properly blown, a paste of any desired thickness consisting of a strong solution of sodium acetate and a precipitate of lead sulphate may be obtained. This method often enables us to get strong solutions and avoid evaporation.

CHAPTER X.

EVAPORATION.

(143) The simplest and cheapest evaporation is that brought about by the sun and wind. Illustrations are found in the separation of salt from salt water, as in Solomon City, Kansas (*Univ. Geol. Survey of Kansas*, Vol. VII, p. 68, 1902); Salt Lake City, Utah (*Sc. Amer.*, Vol. CXV, p. 550, 1916), and San Diego Bay, Southern California (*Met. & Chem. Eng.*, **16**, 317, 1917). At the place last named the gross annual evaporation is 60 inches, the precipitation 10 inches, and the net evaporation 50 inches. At Solomon City the gross annual evaporation is much less and the precipitation heavier. The vats are therefore provided with movable covers which shed the rain and are run off in fair weather.

(144) At Nauheim, Dürrenberg, Rodenberg and Schönbeck, Germany, and at Moutiers in France, where somewhat weak salines occur, Gradirhauser consisting of scaffolds filled with thorn wood (*Prunus spinosa*) and built in the direction of the prevailing wind have been in use for many centuries for evaporating the salt water. In good weather about 60 kilos of water are evaporated in 24 hours for every square foot of surface (*Thorpe's Dict.*, Vol. V). At Nauheim 2000 tons of salt are made annually from salt water concentrated in this way.

(145) Brush racks for evaporating a dilute solution of boracic acid have also been in use at Daggett, Cal. (See *The Saline Deposits of California, Sacramento*, 1902, p. 57). High winds, high temperature and low humidity prevail here and the surface evaporation sometimes amounts to an inch in 24 hours.

(146) Fig. 156 shows an arrangement for evaporation of sprayed salt solutions. A considerable part of the evaporated salt deposits at *H* and is removed through side doors not shown. Success with such an apparatus can only be obtained by careful preliminary study of a small model apparatus. In general it may be said that it is much easier to create such a salt dust than to settle it. The motor *A* should be a steam turbine since this admits of accurate speed regulation, and the air should be driven through

ATOMIZING EVAPORATOR.

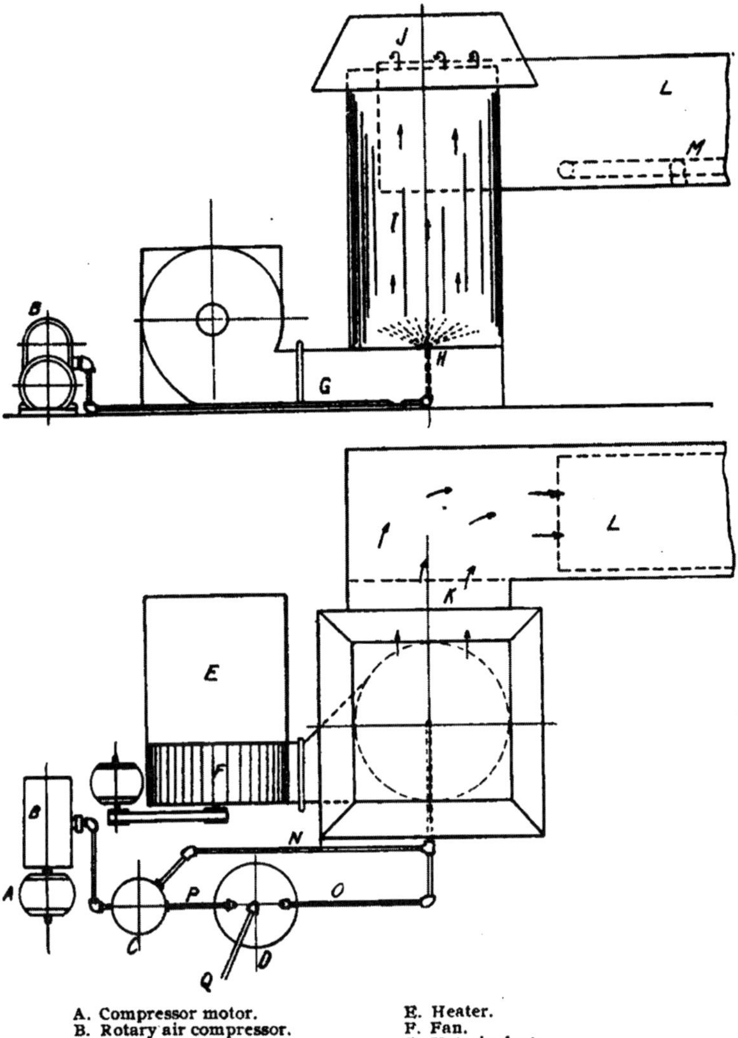

A. Compressor motor.
B. Rotary air compressor.
C. Air pressure tank.
D. Montejus tank for solution.
I. Atomizing tower.
J. Cover for tower.
K. Down shaft to salt chute.
L. Collecting flume,
E. Heater.
F. Fan.
G. Hot air duct.
H. Spray nozzle.
M. Conveyor belt for gathering salts.
N. Air pipe to spray nozzle.
O. Solution pipe to spray nozzle.
P. Air pipe to mounting.
Q. Pipe from solution storage.

Fig. 156.

a fire, the hot gases being used directly after settling, or producer gas or oil may be burned.

(147) The same idea may be used, minus the spray, in a tower filled with chemical brick checker work, the hot gas passing up and the liquid down the tower.

(148) The simplest direct evaporator is a cast iron pot supported over a fire, one or more pots being used over each fire. It is better always to have a brick combustion chamber and then lead the fully burned gases under the pots, since metal surfaces chill the gases and give incomplete combustion. Instead of pots we may use shallow rectangular pans made of sheet iron and the hot fire gases may be lead over the surface or first under the pan and then over the surface of the liquid to be evaporated. Large flues like those of a Cornish boiler may also be built into the evaporating pan. Where direct heating of this sort can be used much less fuel is needed but the efficiency of the fuel, obtained in this way, even under the most favorable conditions seldom exceeds 80 per cent. of theory.

(149) In many cases the evaporation must be carried out in lead lined vessels further reducing the efficiency and often preventing the use of top heating. A serious difficulty often encountered is the formation of adhering crusts which are troublesome to break off and cause heavy additional expense for operation and repairs. For these reasons it is often necessary to resort to indirect heating by means of steam lead in pipes through the liquid to be evaporated or through steam jackets surrounding the vessels.

(150) The best results that can be expected from direct heating are about eight water evaporated to one of fuel used, but nothing like this efficiency is usually obtained. Thus in the salt manufacture as carried on in 1892 (See *Mineral Industry* for that year, p. 418), the consumption of fuel per barrel of salt (280 pounds) was 1/5 to 1/15 ton. If we allow three parts water to one salt we get an evaporation of 3-1 to 5½-1. In steam boiler practice it is customary to call 10 square feet of heating surface the equivalent of 1 horse-power, which under normal conditions is supposed to be able to evaporate 34.5 pounds of water per hour from and at 100° C.

(151) The evaporating capacity of any apparatus depends upon a number of factors. In ordinary evaporation from a surface exposed to the air, evaporation is much more rapid if a current of air passes over the liquid, and the drier the air, of course, the more rapid the evaporation. Where the liquid to be evaporated is not corrosive, the best practice is to heat it to the boiling point in boilers or pipe manifolds placed next to the stacks and discharge from the boilers or manifolds into open evaporating pans or upon revolving drums with scrapers. The particular practice to be followed must depend altogether upon the properties of the substance in solution.

(152) The arrangement of the pans for the finishing operation is best in cascade, the liquid passing from the upper pan by overflow spouts into those lower down and the finished salt being lifted from the lowest pan into a drainer or run into crystallizers. Since this apparatus is liable to develop leakage it is imperative that all parts be easily accessible for repairs. Cast iron pans or pots suitable for such evaporation are made for caustic or other liquids, and for muriatic acid pans up to 12 feet in diameter. For high temperature work they should be supported by the rim so that they may be free to expand and contract. A supporting central pedestal is sometimes necessary when the kettles are very deep. The making of such large castings should be entrusted only to experienced hands. These pots are also supplied with pot and jackets cast in one piece.

(153) For shallow pans of sheet iron or steel, the car building works have machinery capable of pressing them of any desired size and shape. Smaller shops can build them out of sheet metal by cutting square pieces from the corners and bending up the sides and melting the edges together with oxyacetylene torch.

(154) Steam Heated Seamless Cast Iron Pans are made by the Sowers Mfg. Co., of Buffalo, N. Y. Fig. 157 shows such a kettle with legs made in sizes of 5 to 125 gallons. The same kettle cast with lugs may be had in sizes of 150-800 gallons. These

Fig. 157.

Fig. 158.

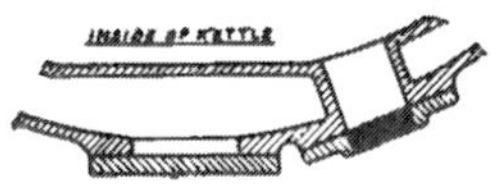

Fig. 159.

Fig. 160.

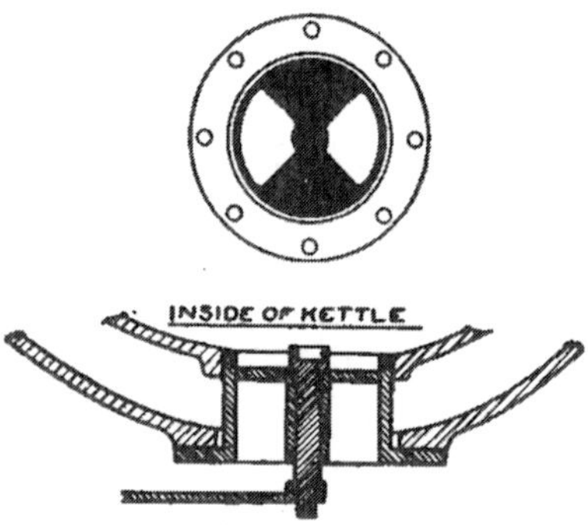

Fig. 161.

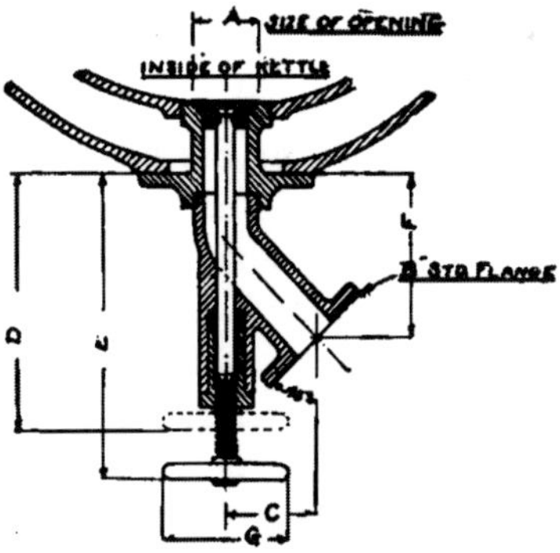

A	B	C	D	E	F	G
2"	2"	4¾	11¼"	13¾	6¾	5"
3"	3"	4½	12¾	15¾	8"	5"
4"	4"	5½	14½	17½	9"	10"
5"	5"	6"	15¾	19"	10"	10"
7"	7	7"	18¾	22"	11½	10"

Fig. 162.

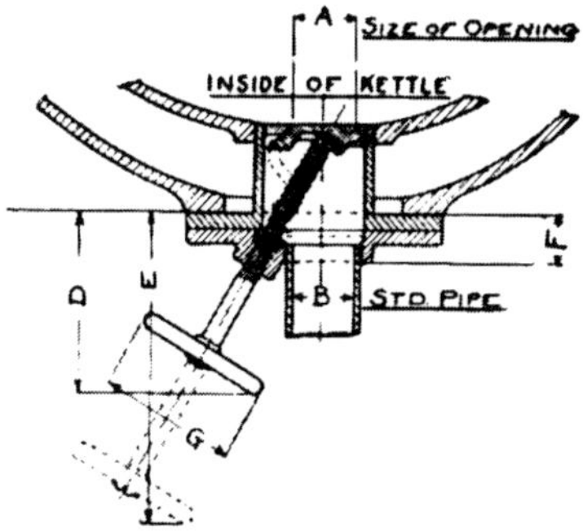

A	B	C	D	E	F	G
2"	2"		6¾	9⅞	2½	5"
3"	3"		7⅜	10 11/16	3"	5"
4"	4"		7½	12½	4¼	5"
5"	5"		13¼	19"	5¼	10
7"	7"		13½	20½	7¼	10

Fig. 163.

Fig. 164.

kettles are fitted with outlets to suit different uses. In Fig. 158 a flanged sleeve is screwed into the inner shell of the kettle and the joint between the bottom of the kettle and the top of the outlet flange is machined to fit tightly at 150 pounds pressure. In Fig. 159 the outlet is necessarily placed to one side of the center. Fig. 160 outlet is Fig. 158 with boss on bottom to which a gate is fitted. This is not suitable for thin liquids. Fig. 161 has a butterfly valve flush with inside of kettle and is used only in kettles equipped with bridge type mixers. Figs. 162 and 163 show other styles to be had.

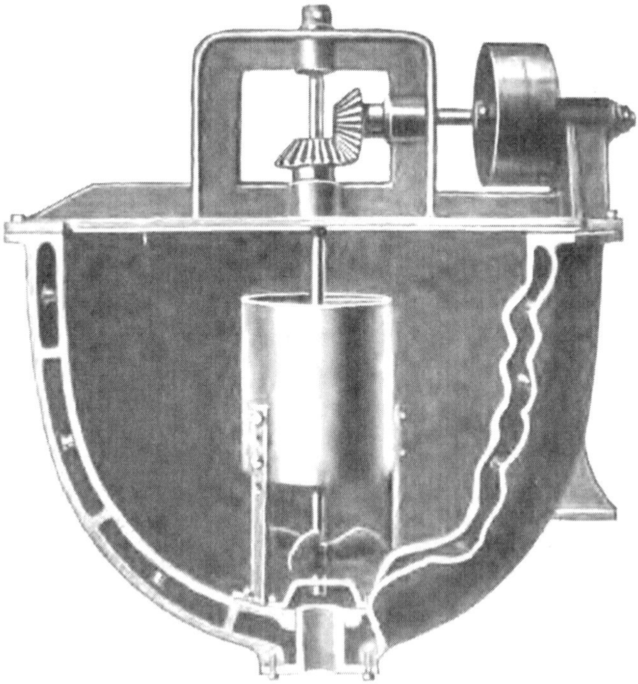

Fig. 165.

Fig. 164 shows a bracket type agitator, and Fig. 165 a propeller screw agitator. Fig. 166 shows a 100-gallon standard type vacuum pan and Fig. 167 a 200-gallon vacuum pan with overhead drive, sweep agitator save all and jet condenser pump.

(155) Vacuum Evaporator.—"If steam at five pounds pressure is introduced into the steam tubes of an evaporator in which the liquid is allowed to boil at the atmospheric pressure the temperature difference between this steam and the boiling liquid (if the latter has the boiling point of water) is about 16° F. If, however, a vacuum of 25 inches is maintained above the liquid to be evaporated the temperature difference is about 95° F. Thus for a given vessel six times the evaporation can be accomplished in a given time, or for a given amount of evaporation the heating surface in the vessel may be one-sixth the size." (*J. Franklin Inst.*, **166**, p. 249, 1908). With a Dopp cast iron 300-gallon vacuum pan

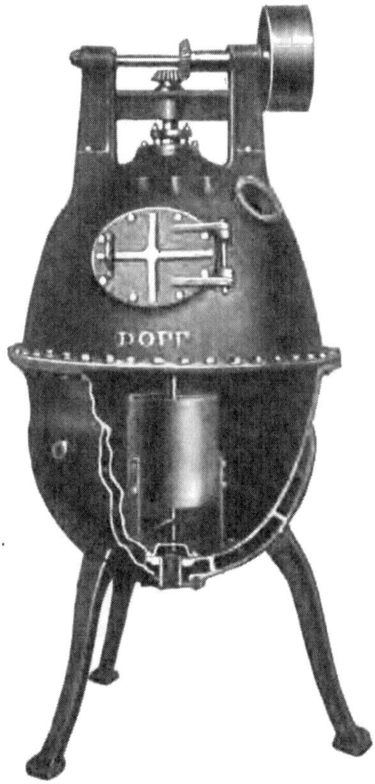

Fig. 166.

with heating surface of 17.34 square feet the following evaporation was obtained in an hour:

FEED AT TEMPERATURE OF VACUUM.

Vacuum	Steam pressure in jacket 80 lbs.	120 lbs.
24 in.	44.37 gal.	51.59 gal.
26 in.	49.50 gal.	56.89 gal.

Fig. 167.

Where copper or brass is the heating surface a higher rate prevails and with lead a lower. Thus with round copper tubes into which the steam is led the coefficient of transmission of heat in calories per square meter per hour for steam and water is $K = \frac{1900}{dl}$, where d is the diameter of the tube and l the length in meters. For wrought iron pipe this value is multiplied by 0.75, for cast iron and for lead 0.45 (Viola, *Met. & Chem. Eng., X*, 31, 1912).

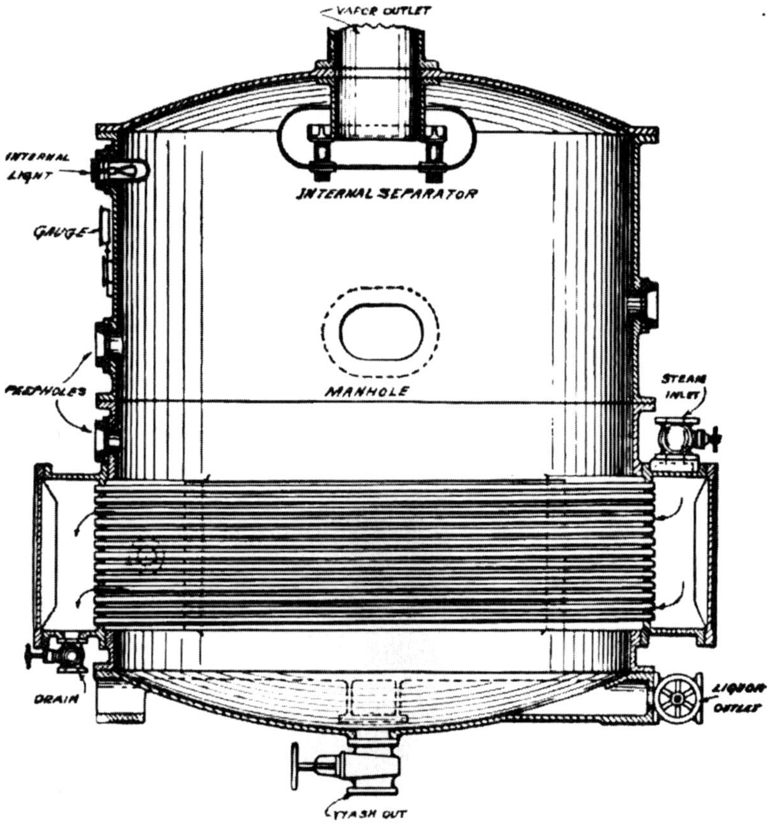

Fig. 168.

(156) A single effect vacuum evaporator such as the Zaremba evaporator, shown in Fig. 168, has essentially the same construction as a tubular boiler except that steam is passed through the

Fig. 169.

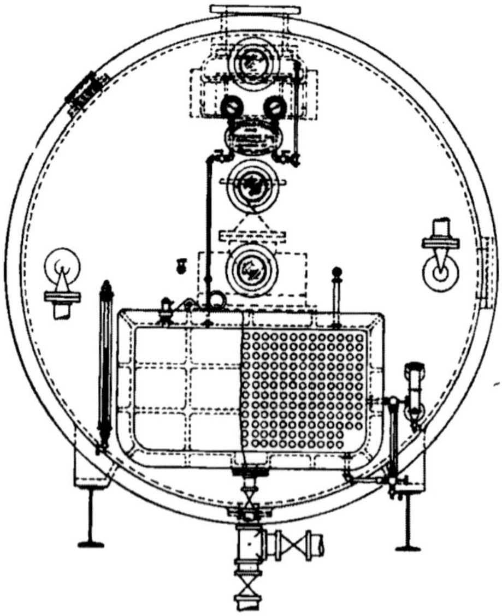

Fig. 170.

tubes instead of fire gases. The Buffalo Foundry & Machine Co., build a horizontal tube evaporator shown in Figs. 169, 170 and 171. This is intended especially for the evaporation of non-foaming liquids and is made in two sizes 96 inches and 114 inches diameter with a range of heating surface from 590-2460 square feet. For concentrating liquids, such as caustic soda to high densities they also furnish a special form of evaporator shown in Fig. 172. For caustic the apparatus, including the tubes, is made entirely of cast iron. The tubes are 4 inches in diameter and are placed close together. Steam is supplied by small tubes

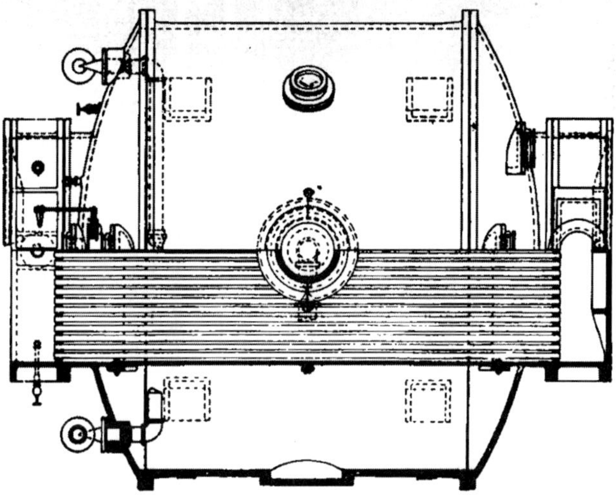

Fig. 171.

placed inside reaching to their ends. Care has been taken to reduce the number of joints and to protect them by special gaskets. This is built in four sizes with 110-660 square feet radiating surface.

(157) Multiple Effect.—"Suppose we had three single effect evaporators *A*, *B* and *C*, of the same size, each with an independent surface condenser and air pump. *A* takes exhaust steam at a temperature of 225° and delivers vapor to condenser *A* at 195°. *B* takes steam at 195° and delivers its vapor to condenser *B* at 165°. *C* takes steam at 165° and delivers vapor to condenser

C at 125° (Fig. 173). In this case excluding radiation and other losses, we would require for each pound of evaporation 1 pound of steam for producing the vapor, and about 25 pounds of cool

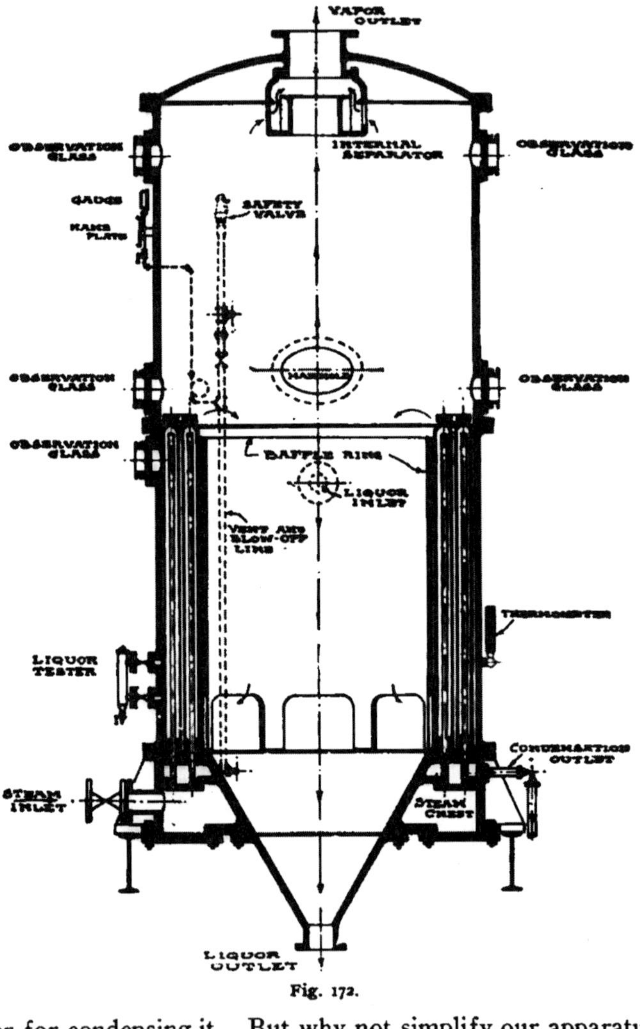

Fig. 172.

water for condensing it. But why not simplify our apparatus by rearrangement as follows: Eliminate condenser *A* and substitute for it the heating surface of single effect *B*—that is, let the vapor

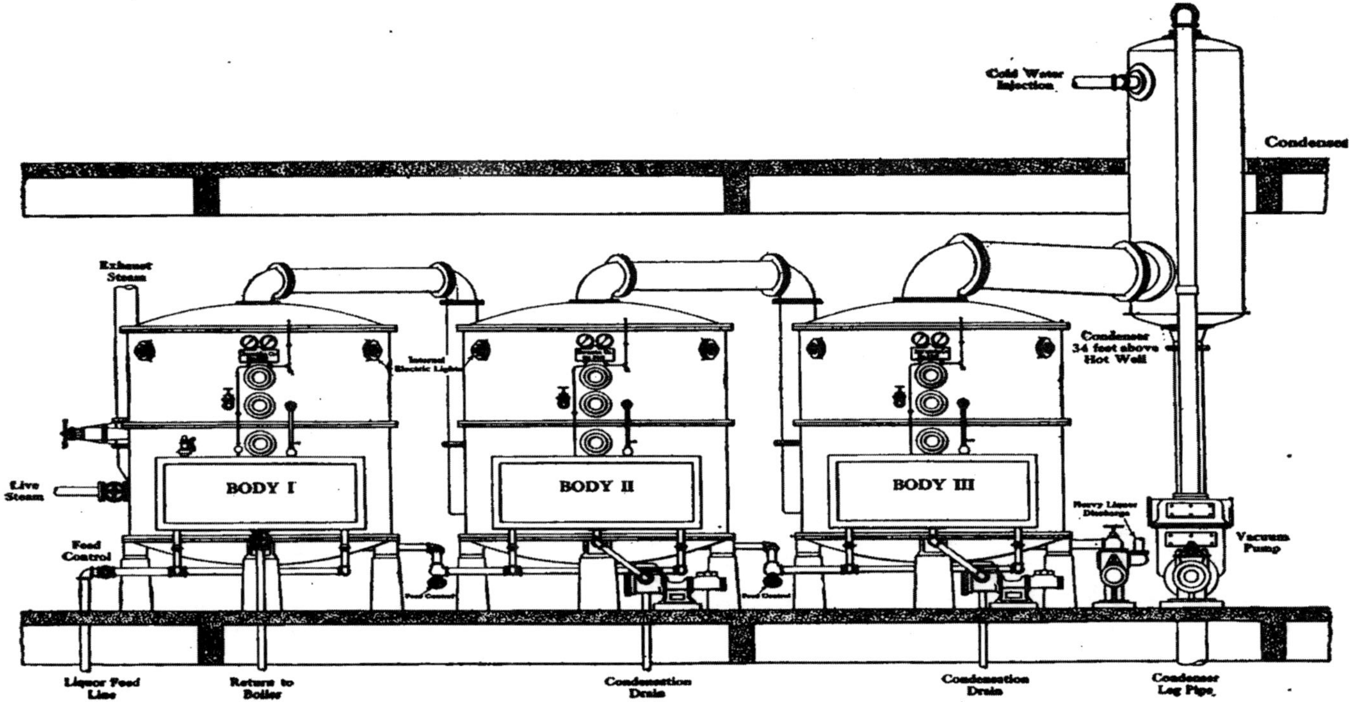

Fig. 173. Multiple Effect.

from single effect *A*, take the place of the 195°-steam previously fed into single effect *B*. In like manner let the vapor from single effect *B* be condensed in the tube of single effect *C*, while the vapor from the latter is condensed in condenser *C* as before. With our new method of operation the temperature difference in each effect is the same as before, so also is the amount of work done, but we are feeding steam into effect *A* only, and are injecting cold water into condenser *C* only, and are using but one-third as much of each as in the first case. Furthermore, we have discarded condensers *A* and *B*, together with their respective vacuum pumps. In the new dispensation we are working under the multiple effect system—that is, we have converted our three single effects into one triple effect thereby increasing our evaporation to 3 pounds of water per pound of steam used, and decreasing our water consumption from 25 pounds to about 9 pounds per pound of evaporation, the amount of evaporation produced being practically the same in both cases. The theoretic steam economy of the multiple effect, as stated above, is modified by a number of conditions, chief among which is the initial temperature of the liquor fed and the disposition made of the heat contained in the condensation from the various bodies. Ordinarily the number of effects used does not exceed four, but for special purposes as many as seven or eight effects can be used advantageously; it all depends upon the circumstances."

"The pumps necessary for operating the multiple effect consist of a vacuum pump for removing the air and other noncondensable gases from the condenser, a tail pump to withdraw the finished liquor from the evaporator, and condensation pumps to draw away the condensation from the steam chests operating at less than atmospheric pressure. As the steam exhaust from these pumps is used in the evaporator their operation costs practically nothing. In such circumstances to use motor driven pumps means a loss in economy."

"All forms of vacuum apparatus call for the use of a condenser in which the vapor coming from the last pan is condensed by contact with cold injection water. In the wet system the air and gases present are entrapped in this warm mixture of condensed vapor and injection, the entire mass being withdrawn from the system by a wet vacuum pump discharge. In the dry system of condensation the condenser is made much larger and is mounted sufficiently high (about 34 feet) to allow the water discharge to flow into the hot well by its own weight against the pressure of the atmosphere. In this case air and non-condensable gases only are handled by the vacuum pump. In the surface condenser, which is practically a reversed evaporator, the vapor is confined to one chamber and the cooling water to another, being separated from each other by a tubular brass cooling surface."

"As a rule the amount of water required varies from 25 to 30 pounds per pound of vapor condensed. One of the great advantages of the wet system lies in the fact that such a condenser can generally be made to draw its own supply of injection water from a neighboring stream or tank. The dry system requires a circulating pump. There is a large supply of warm water at about 110° F. coming from the condenser, which can frequently be used to very good advantage. The first pan delivers a quantity of distilled water at 212°, equal to the amount of steam used; this is preferably returned to the boiler. From the condensation pumps another supply of warm water is obtained at a temperature between the two former."

"The steam pressures ordinarily carried vary from 5 pounds down to atmospheric pressure; the vacuum from 23 to 26 inches."

"There are two ways of handling the concentrated liquor. One method, "batching," calls for the carrying of the concentration of the entire body of liquor in the final pan to the proper finishing density, after which the heavy liquor is removed by the tail (or magma) pump. In the "continuous" method the heavy liquor settling in the bottom of the final pan is withdrawn by the pump while the pan continues in operation." (Zaremba Co., Catalog, 1916.)

(158) It is evident that the heat in the multiple effect does double to quadruple work according to the number of effects. Theoretically it should be possible to use the heat over and over again indefinitely, but in practice a limit is found, due in part to the loss of heat by radiation and in part to the increased cost and complexity of the apparatus. "In the salt vacuum pan the evaporation per pound of coal guaranteed is 6 pounds in the single effect, 12 pounds in the double effect, 18 pounds in the triple effect and 24 pounds in the quadruple effect from brine of 95 salinometer." (Letter of July 16, 1918, Manistee Iron Works, Manistee, Mich.)

(159) It is best that each pan have its own vacuum pump and that each multiple effect have its own condenser. In calculating, the evaporative power of the coal for raising steam for the pumps and for pumping water must be added to that as steam supplied direct to the first vacuum pan.

Fig. 174.

(160) The Zaremba Crystallizing Evaporator shown in Figs. 174-176 is designed for the evaporation of solutions that tend to encrust the heating surface. The circulation of liquid is very rapid and the tubes may be readily cleaned mechanically. Fig. 174 shows the heating surface. The circulation is up through the tubes and down through an annular space between the steam chest and the outer shell. The separated crystals drop into the cone below the steam chest from which they drop into the salt filter and are removed through the door in front. Fig. 176 shows this apparatus in double effect. The construction is such that

Fig. 175.

the liquid in rising has a velocity of 30 feet per second while because of the greater area of the annular space the drop has a velocity of only one-tenth this so that the crystals readily settle

Fig. 176.

out. The salt filter has a sight glass and when filled with crystals these are washed with weak liquor by closing and opening appropriate cocks and the crystals are then removed. Any number of effects may be used and are so connected that one may be cut out without interfering with the others. As each effect operates at a different temperature it is possible to effect many separations.

(161) In the Lillie evaporator the steam tubes are horizontal and the liquid to be evaporated is showered over them. In the Kestner evaporator the liquid runs into upright tubes 23 feet long surrounded by steam. Bubbles of steam form and mount the tubes carrying a thin film of liquid which evaporates rapidly. The foam at the top strikes against the vanes of a centrifugal separator which separates steam and liquid completely. The liquid evaporates with tremendous rapidity and is exposed to the heat for so short a time as not to injure organic dissolved material, such as cane sugar, citric acid, tartaric acid and glue. Any deposit, such as gypsum, is blown out and settles in the pans. The Yaryan apparatus has horizontal tubes for film evaporation.

(162) Acid Linings.—Fig. 177 shows a method of constructing acid proof linings for vacuum evaporators used by the Zaremba Co. The difficulty of evaporating acid liquids by multiple effect lies chiefly in the difficulty of heating. If metal pipes are used they are quickly corroded. If enameled pipes are used they are very poor conductors of heat. For some acid liquids containing nitric or sulphuric acid duriron or one of the other iron silicon alloys may be used but these alloys are very hard and very brittle.

(163) By using steam at 100 lbs. pressure in an evaporator constructed of steam pipes opening into the steam chest at the bottom and closed at the top the Zaremba Company have been able to concentrate caustic soda and other liquids of high boiling point so that solidification takes place on cooling.

(164) Another form of multiple effect, the Söderlund-Boberg evaporator, has recently appeared which if the statements made in its behalf are verified bids fair to revolutionize this branch of the industry. The apparatus is shown in single effect in Fig. 178

The evaporator comprises a tall shell *A*, the greater portion of which is taken up by the calandria *B* to the upper end of each tube *C*, of which is fitted a ferrule with a distributing nozzle *E*. The liquid which is raised by a pump *F* from the lower portion *G* of the shell *A* is fed into the tray *H* formed at the top of the calendria *B*, and passing through the nozzles *E* in the ferrules *D*, is spread as a thin and rapidly moving sheet over the inner surface of the tubes *C*.

Fig. 177.

The liquid collects again in the bottom of the shell, to which fresh liquid is automatically admitted by a constant lever feed *I*, to be returned by the pump *F* to the distributing tray *H*, while the stream given off by the films in the inside of the tubes and sep-

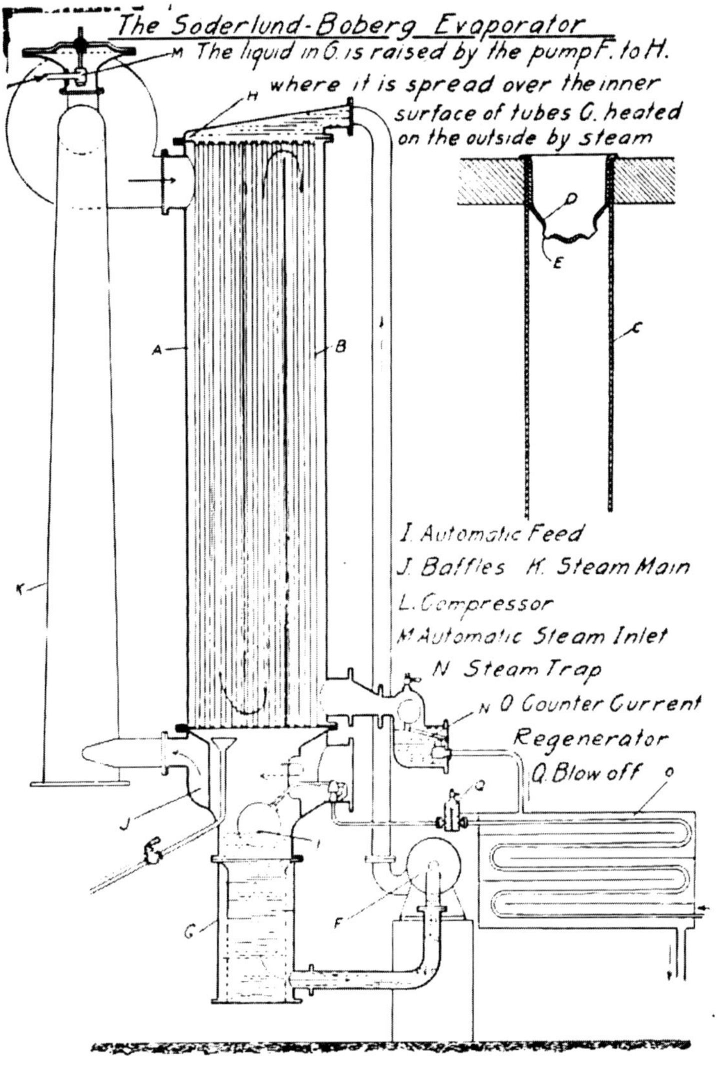

Fig. 178.

arated from the spray by the baffles *J* in the base of the shell and by the tangential disposition of the steam entry to the steam main *K* leading to the compressor *L* passes to the compressor, having the required quantity of make-up steam added to it on its way by a pressure controlled automatic steam inlet *M*. The compressed steam enters the space around the tubes *C* and condensing on them is drained off as hot water through a steam trap *N* connected to a counter current regenerator *O* through which the cold feed flows to the automatic feed valve *I*. The small quantity of non-condensable gases which is not driven off from the feed liquid through a blow-off *Q* at the hot end of the regenerator *O* is cleared out of the steam space by periodically blowing off a little steam or continuously allowing a small leak of steam out of this space. The compressor is driven by a constant speed by spur gear from a shaft belt driven direct from a gas engine, working for instance in connection with a small suction gas plant or by an oil engine. The exhaust in either case passes to a small boiler generating the make up steam.

In this apparatus, it is asserted, 240 pounds of distilled water may be obtained for every pound of coal burned.[1]

(165) Evaporation of Acid Liquids.—Liquids containing free hydrochlorine acid must be evaporated in glass, glass-enameled, stoneware or silica vessels. This may be done by blowing heated air or superheated steam through the liquid or by allowing the liquid to flow down and over checkerwork in a tower up which the heated gases are passing. Another method is to use glass or silica tubes as shown in the Hart Evaporator in Fig. 3. In this figure l is a longitudinal section, 2 plan and 3 a cross section. (See *Jour. Ind. & Eng. Chem.*, July, 1918, Vol. X, p. 555.)

[1] A discussion of the theory and practice of multiple effect evaporation will be found in the *Journ. Ind. & Eng. Chem.* for May, 1918, Vol. X, p. 191.

CHAPTER XI.

CRYSTALLIZATION.

(166) Most solids are more soluble in hot than in cold liquids and separate out as crystals from such solutions on cooling. Because of a difference in solubilities this process often offers a ready and cheap method of purification. As an illustration we may take the preparation of potassium nitrate from Chile saltpeter ($NaNO_3$) and potassium chloride (KCl). When these are dissolved in hot water to saturation in the proportion:

$$NaNO_3 + KCl = 85 : 74.6,$$

we have four substances: $NaNO_3$, KNO_3, $NaCl$, KCl in solution. If we evaporate such a solution we soon reach a point where the common salt can remain in solution no longer, because common salt is very little more soluble in hot than in cold water, and it begins to crystallize out in cubes. On the contrary the three other salts are much more soluble in hot water than in cold and they remain in solution. As the common salt separates more forms in solution. The $NaNO_3$ and KCl diminish in quantity and the KNO_3 increases, salt continually separating, until the solution becomes saturated with KNO_3 and it, too, begins to crystallize out. The boiling is now stopped and the solution run into crystallizers. As the temperature falls in the crystallizers much potassium nitrate and a little salt separate.

(167) In order that the crystals of salt separated may be purified from adhering potassium nitrate solution, it is necessary that the crystals shall be small and if the boiling is rapid this will be the result. As the salt forms it is lifted out with perforated shovels thrown into a centrifugal, whizzed, to throw out the mother liquor, washed in the centrifugal with a little cold water and dried. This salt still contains a small quantity of nitrate which does not, however, interfere with its use for most purposes. In order to get it perfectly pure we should have to dissolve it and boil it down again, or wash it in the counter current apparatus.

(168) The crystals of potassium nitrate which begin to form in the crystallizers will become very large if the solution cools slowly and quietly and films and globules of mother liquor will be enclosed in these large crystals. To prevent the formation of large crystals the solution is stirred as it cools and when cooled perfectly the crystals are whizzed in a centrifugal and rotated in a macerator with a cold saturated solution of potassium nitrate. The nitrate will not dissolve in this solution but the salt will and on again whizzing it in the centrifugal and washing with a little more solution of pure nitrate we get a nearly pure potassium nitrate. If this must be absolutely free from salt it must be redissolved in boiling water and recrystallized.

(169) It becomes evident from the above that means for controlling the size of the crystals is very desirable in order to insure greater purity. Too little attention has been paid to this point, except in the crystallization of cane sugar. When the cane juice has been properly evaporated in a multiple effect it is transferred to a single effect with high vacuum and the boiling continued.

(170) "As the evaporation goes on, the liquid becomes thicker and thicker until a density is reached at which the sugar, under the existing temperature in the pan, begins to crystallize. It is now that close attention is necessary in order that at the proper moment crystallization shall be checked, or rather, that the formation of an undue number of crystals shall be prevented so that finally crystals of the desired size shall be obtained. As the critical moment approaches, the pan-boiler takes samples of the boiling liquor out of the pan by means of a proof stick. This is a metal rod about 1 inch in diameter and about 2.5 feet long, freely running in a sleeve in the side of the pan. The sleeve has a ground face against which a ground shoulder on the proof stick fits in such a manner that a fairly tight contact is made and air is prevented from leaking into the pan. In the pan end of the proof stick there is a cup like depression in which about an ounce of liquid can be drawn from the pan. As the pan man draws this out, he throws the liquid on to a glass plate, or allows it to flow between his finger and thumb, at the same time examining it before a bright light. If the density of crystallization has been reached, the first young crystals will appear as tiny points of light flowing down the glass, and it will be necessary to watch the rate of their increase very closely, for it is very rapid. The moment that the operator judges that enough of these crystals have been formed, he quickly opens the valve communicating with a suction tank from which a supply of liquor may be drawn in; the inflowing thinner liquor brings the boiling mass to a density at which no more crystals can be formed, but those remaining in the pan will begin to grow in size as the density of the liquor again begins to rise. It is, now that the nursing of the crystals goes on, and in order that they shall grow in vigor and in strength, they are treated to occasional "drinks," *i. e.*, liquor is drawn in whenever the density of the mother liquor tends to rise to the point at which a new lot of crystals would be born; if this last should happen, trouble would result for this second crop coming into the world so much later than the first would not grow to the same size as their older brethren, and thus the resulting sugar would show an uneven grain when finished and the washing in the centrifugal would be difficult. When the boiling mass, which is now a mixture of crystals floating in the mother liquor (syrup), through repeated "drinks," fills the pan, it is finally brought to a proper consistency, and the boiling is finished. The vacuum is broken, the steam is turned off and the whole mass allowed to flow through a foot-valve into a "mixer," where it is kept in motion by means of revolving paddles, and is drawn off in batches to the centrifugals to be purged from the syrup. The last traces of syrup that tend to adhere to the now white crystals on the walls of the centrifugals are washed away by spraying distilled or filtered water into the machines. The centrifugals are allowed to spin until the sugar is fairly dry, this last is then conveyed to a granulator to be dried." (*Chemical Industry in Canada*, p. 40, Toronto, 1909.)

(171) This growing of crystals is possible when there is not too much difference in the specific gravities of solid and solution. The sugar in this case has sp. gr. 1.593 while the syrup is about 1.33. In the case of ferrous sulphate these figures are 1.89 and 1.24 and in the case of common salt 2.16 and 1.21. As the difference grows some other means must be used to establish and maintain this control. If the solution is run into a cylinder which can be rotated on its axis and is cooled while rotating by running cold water over the outside the crystals will form rapidly and will be small so that they can be easily freed from mother liquor by very little washing. Treated in this way alum, potassium nitrate and sulphate, barium nitrate and many other solutions can be treated very successfully but some other solutions are apt to form a thick mush of very fine needles difficult to handle.

(172) It has been suggested[1] that the viscosity of a liquid probably has also much to do with the growth of crystals in suspension. Whatever the cause may be the control of crystallization and its rapid completion so as to avoid the building of expensive crystallizing pans is greatly to be desired.[2]

(173) Crystallizing Pans are usually made of wood, sheet steel or cement. The first is commonly cheaper. If case wood is used the pans will swell and shrink with moisture and temperature changes. They are sure to leak and must be placed over a tight floor inclined towards a sump and pump. Wooden crystallizing tanks must be shallow since wood is a poor conductor and most cooling is from the surface. Sheet steel crystallizers may have a hopper shape with a door in the bottom so that the crystals can be shoveled directly into barrels or into a centrifugal. Cement crystallizers are stronger when cylindrical. For further discussion see under "Tanks."

[1] By F. C. Zeisberg in a letter to the author.

[2] Vant Hoff has studied the interactions of 4 salts in solution (*Vorlesungen in Theoret. u. phys. Chem.*, 1898, p. 90) and the results have been applied to the utilization of salt water mother liquor by Hildebrand. (*J. Ind. Eng. Chem.*, 10, 96, 1918.)

CHAPTER XII.

DRYING.

(174) Tunnel dryers are sometimes used for drying small quantities of materials. They consist of closed compartments with steam heating pipes and ventilating fans for renewing the air, and muslin screens for removing dust and soot from the

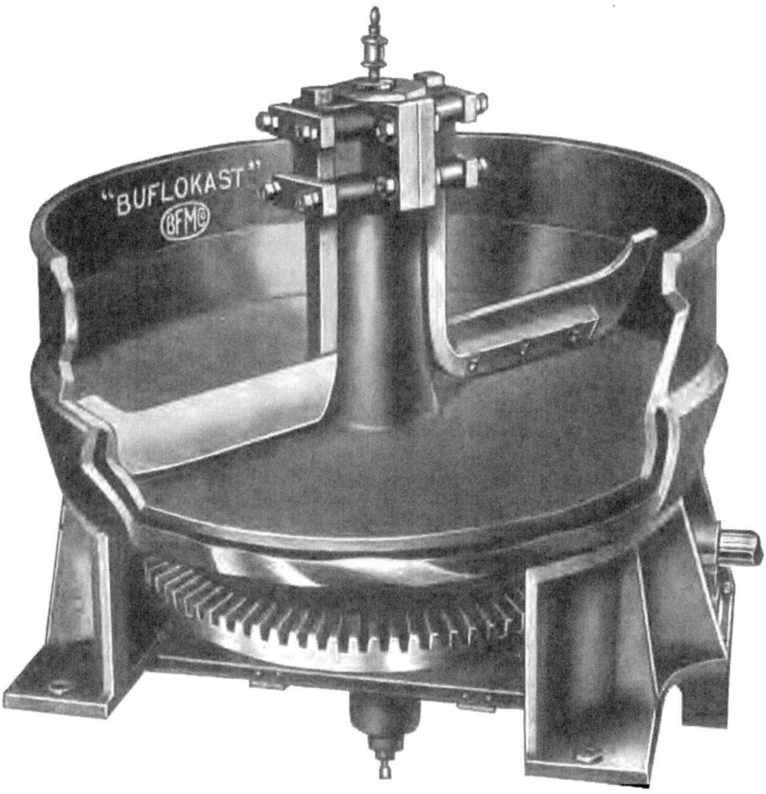

Fig. 179.—Crystallizing and Drying Pan.

entering air when necessary. The material is placed on frames covered with cloth supported on racks placed on cars which are run into one and out the other end of such a tunnel. Or racks are placed on the sides of the tunnel, the frames filled, placed in the tunnel and when dry removed, emptied, refilled and replaced. Some substances, like citrate of iron and ammonia, are dried in scales by pouring on glass plates and when dry scraping off with

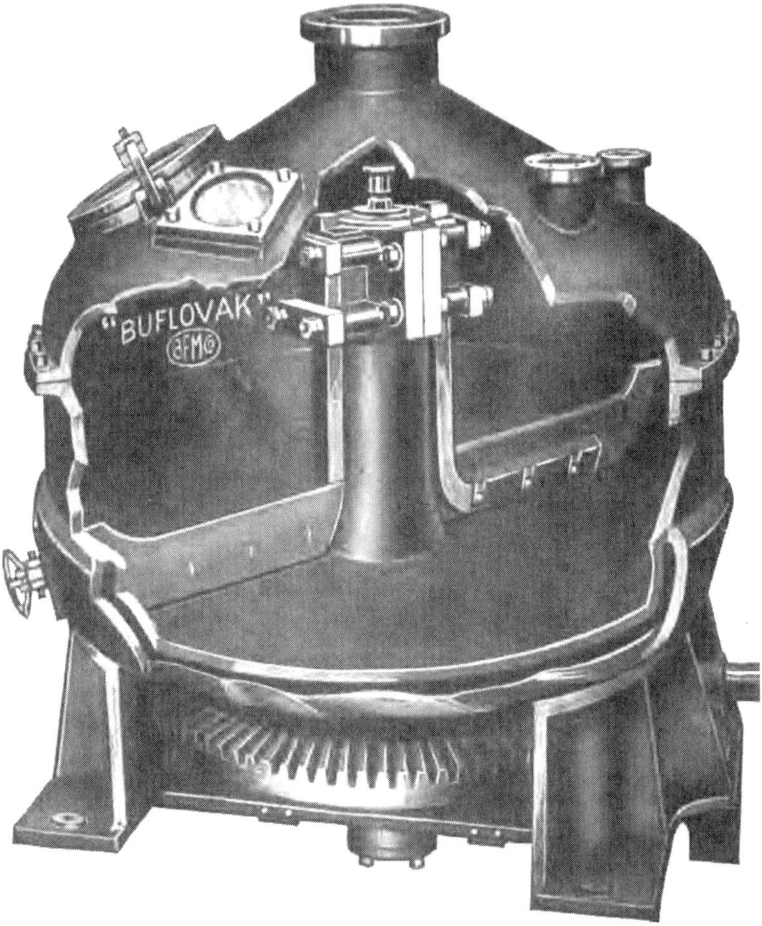

Fig. 180.—Vacuum Crystallizing and Drying Pan.

a knife. These shelf dryers may be made of metal tightly bolted so as to be air tight. If a vacuum is produced in them rapid drying is secured at a lower temperature.

(175) For evaporating and drying substances like ammonium nitrate the pan shown in Fig. 179, made by the Buffalo Foundry & Machine Co., is used. For evaporation, steam is injected into the cast iron jacket and when evaporation has proceeded far enough, cold water is run through the jacket to cause the mass to crystallize. This pan may also be used as a dryer for substances like T. N. T.

(176) The form shown in Fig. 180 made by the same company is used for vacuum work, where a low temperature is necessary.

A vacuum shelf dryer made by the same company is shown in Fig. 181. This cut also shows the pump and condenser.

(177) Fig. 182 shows a direct heat, shelf retort made by the same company for reclaiming high boiling point solvents.

It is provided with ducts passing from one side to the other. Shelves are formed between the ducts. The retort may be operated under vacuum. Hot gases from the furnace pass through the ducts and the temperature of each shelf is regulated by a separate damper.

(178) It has been known for many years that the pig iron output of an iron blast furnace varied considerably with the weather conditions and that wet weather was sure to cause disturbances. Mr. James Gayley showed[1] that it was often profitable to remove most of the moisture by cooling the air and then reheating it so as to secure constant moisture conditions in the furnace. Mr. Gayley's experiments at the Isabella Furnaces were conducted by passing the air over pipes containing brine cooled by an ice machine. The coils were soon coated with ice, which is a poor conductor, resulting in decreased efficiency and cooling the air to a temperature lower than was necessary.

In many drying operations it is very desirable to have dry air of uniform low moisture content. Some operations such as the handling of silk require uniform temperature and moisture. This regulation of moisture is spoken of as conditioning.

(179) The Carrier Air Conditioning Co., have shown that by blowing a spray of water at 35° through air and then passing

[1] *Trans. Am. Inst. Min. Eng.*, 35, 746, 1905.

Fig. 181.—Vacuum Shelf Dryer with Pump and Condenser.

through a separator, the moisture may be reduced with greatly simplified apparatus, and at less cost since no ice is formed and no loss in conductivity takes place.

Fig. 182.—Direct Heat Shelf Retort.

(180) The late W. B. Ruggles contributed in 1909 to Kent's Mechanical Engineers Pocket Book a very able article on Dryers

and Drying which with the kind permission of the editor I am permitted to reproduce as follows:[1]

"Materials of different physical and chemical properties require different types of drying apparatus. It is therefore necessary to classify materials into groups, as below, and design different machines for each group.

"Group A: Materials which may be heated to a high temperature and are not injured by being in contact with products of combustion. These include cement rock, sand, gravel, granulated slag, clay, marl, chalk, graphite, asbestos, phosphate rock, slaked lime, etc.

"The most simple machine for drying these materials is a single revolving shell with lifting flights on the inside, the shell resting on bearing wheels and having a furnace at one end and a stack or fan at the other. The advantage of this style of machine is its low cost of installation and the small number of parts. The disadvantages are great cost of repairs and excessive fuel consumption, due to radiation and high temperature of the stack gases. If the material is fed from the stack and towards the furnace, the shell near the furnace gets red hot, causing excessive radiation and frequent repairs. Should the feed be reversed the exhaust temperature must be kept above 212° F., or recondensation will take place, wetting the material.

"In order to economize fuel the shell is sometimes supported at the ends and brickwork is erected around the shell, the hot gases passing under the shell and back through it. Although this method is more economical in the use of fuel, the cost of installation and the cost of repairs are greater.

"Group B: Materials such as will not be injured by the products of combustion but cannot be raised to a high temperature on account of driving off water of crystallization, breaking up chemical combinations, or on account of danger from ignition. Included in these are gypsum, fluorspar, iron pyrites, coal, coke, lignite, sawdust, leather scraps, cork chips, tobacco stems, fish scraps, tankage, peat, etc. Some of these materials may be dried in a single-shell dryer and some in a bricked-in machine, but none

[1] Edition of 1913, page 547.

of them in a satisfactory way on account of the difficulty of regulating the temperature and, in some cases, the danger of the explosion of dust.

"Group C: Materials which are not injured by a high temperature but which can not be allowed to come into contact with products of combustion. These are kaolin, ocher and other pigments, fuller's earth, which is to be used in filtering vegetable or animal oils, whiting and similar earthy materials a large portion of which would be lost as dust in direct heat drying. These may be dried by passing through a single shell dryer incased in brickwork and allowing heat to come into contact with the shell only; but this is an uneconomical machine to operate, due to the high temperature of the escaping gases.

"Group D: Organic materials which are used for food either by man or the lower animals, such as grain which has been wet, cotton seed, starch feed, corn germs, brewers' grains, and breakfast foods, which must be dried after cooking. These, of course, can not be brought into contact with the furnace gases and must be kept at a low temperature. For these materials a dryer using either exhaust or live steam is the only practical one. This is generally a revolving shell in which are arranged steam pipes. Care should be exercised in selecting a steam dryer which has perfect and automatic drainage of the pipes. The condensed steam always amounts to more than the water evaporated from the material.

"Group E: Materials which are composed wholly or contain a large proportion of soluble salts, such as nitrate of soda, nitrate of potash, carbonates of soda or potash, chlorates of soda or potash, etc. These in drying form a hard scale which adheres to the shell, and a rotary dryer can not be profitably used on account of frequent stops for cleaning. The only practical machine for such materials is a semicircular cast-iron trough having a shaft through the center carrying paddles that constantly stir up the material and feed it through the dryer. This machine has brick side walls and an exterior furnace; the heat from the furnace passing under the shell and back through the drying material or out through a stack or fan without passing through the material, as may be desired. Should the material also require a low temperature, the same type of dryer can be used by substituting steam jacket steel sections instead of cast iron."

The efficiency of a dryer is the ratio of the theoretical heat required to do the drying to the total heat supplied. The greatest loss is the heat carried out by the exhaust or waste gases; this may be as great as 40 per cent. of the total heat from the fuel, or with a properly designed dryer may be as small as 8 per cent. The radiation from the shell or walls may be as high as 25 per cent. or as low as 4 per cent. The heat carried away by the dried material may amount under conditions of careless operation to as much as 25 per cent. or may be as low as nothing.

A properly designed dryer of the direct heat type for either group "A" or "B" will give an efficiency of from 75 per cent. to 85 per cent.; a bricked-in return draught single-shell dryer, from 60 per cent. to 70 per cent.; and a single shell straight draught dryer, from 45 per cent. to 55 per cent. A properly designed indirect-heat dryer for group "C" will give an efficiency of 50 per cent. to 60 per cent., and a poorly designed one may not give more than 30 per cent. The best designed steam dryer for group "D" in which the losses in the boiler producing the steam must be considered, will not often give an efficiency of more than 42 per cent.; and, while a poorly designed one may have an equal efficiency, its capacity may not be more than one-half of a good dryer of equal size. The dryer described for group "E" will not give an efficiency of more than 55 per cent.

Fig. 183 gives details of a single shell dryer arranged for the drying of nitrate of soda. This machine has a shell 36 inches in diameter and 20 feet long, equipped with a specially designed furnace and has a capacity of 27 tons of nitrate of soda per 24 hours, reducing the moisture from 5 per cent. to less than one-half of 1 per cent. The fuel consumption per ton of nitrate of soda, is 28 pounds of coal with a thermal value of 14000 B. t. u. and the power required is 6 horse-power. One man can operate the machine and do the firing, as well as attending to any feeding and delivering conveyers.

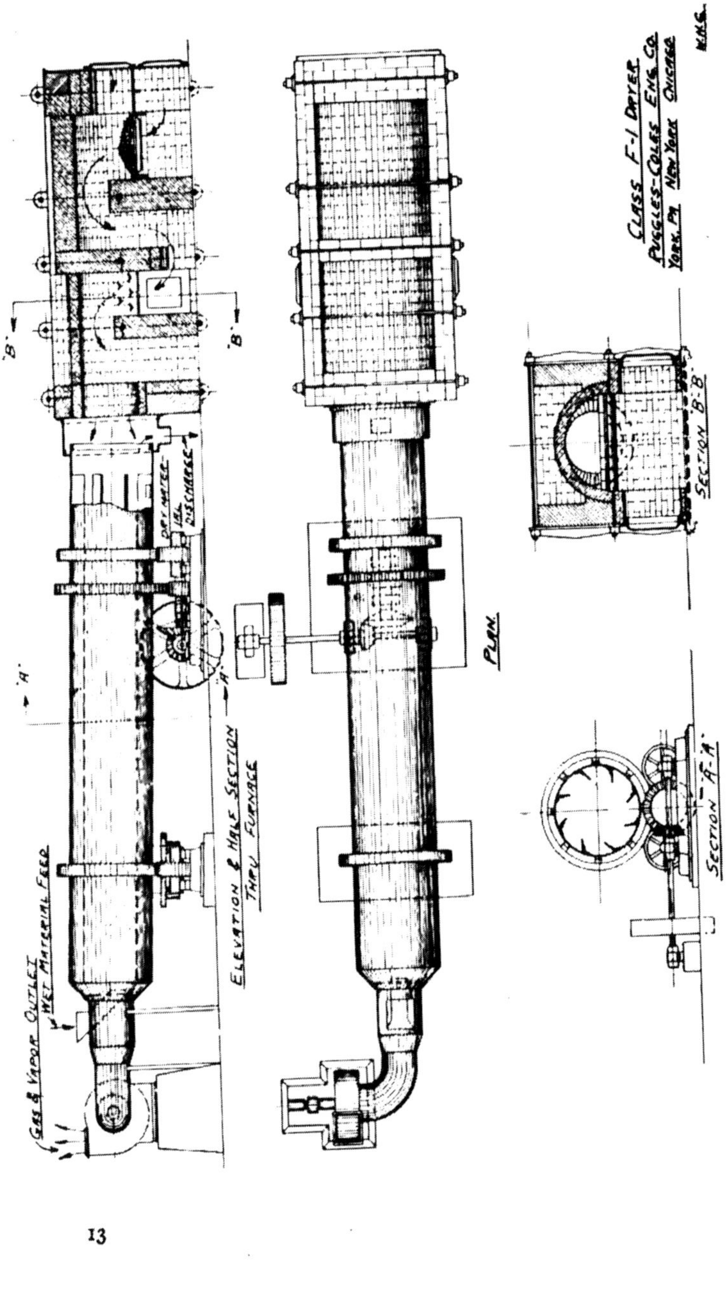
CLASS F-1 DRYER
RUGGLES-COLES ENG. CO.
YORK, PA NEW YORK CHICAGO
PLAN
ELEVATION & HALF SECTION
THRU FURNACE
SECTION 'B-B'
SECTION 'A-A'
GAS & VAPOR OUTLET
WET MATERIAL FEED
DRY MATERIAL DISCHARGE

Fig. 183.

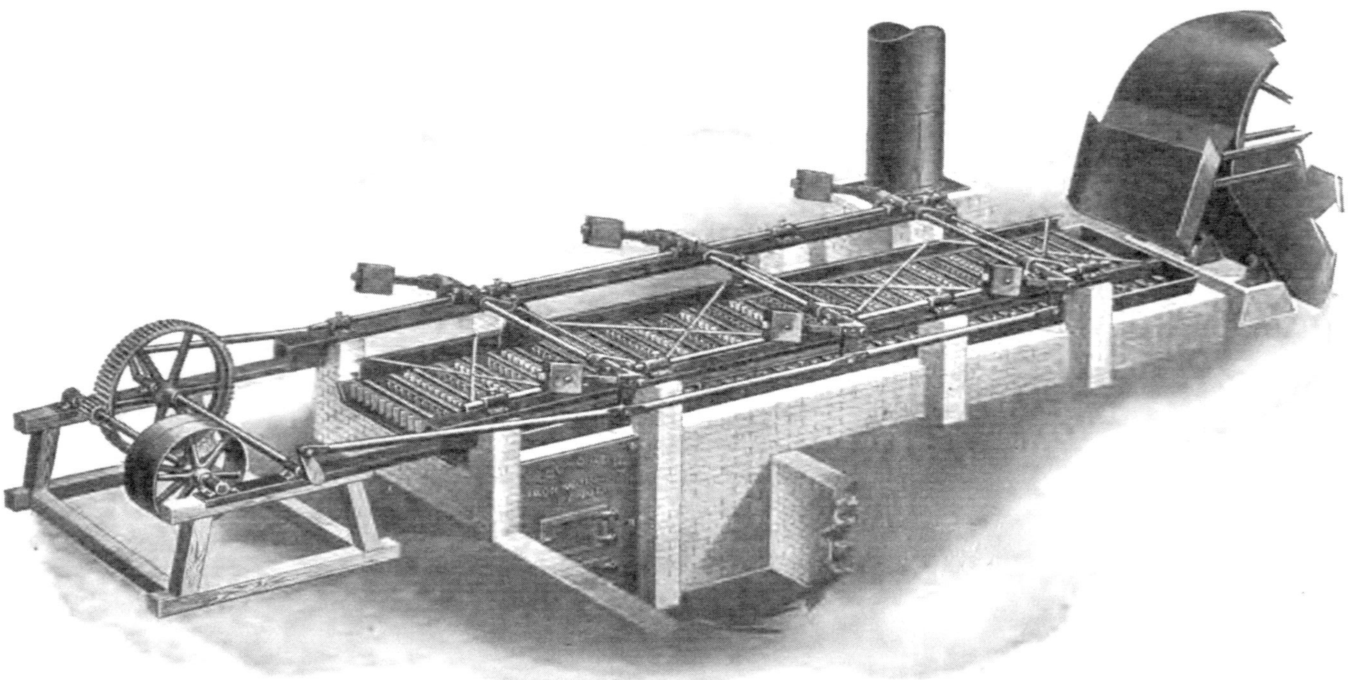

Fig. 184.

(181) **The Lowden Dryer** manufactured by the Colorado Iron Works Co., Denver, Colo., is shown in Figs. 184, 185 and 186.

"The drying is done upon a hearth which is composed of cast iron plates, *A*, beneath which is the flue *B*, through which the gases of combustion pass. The plates comprising the hearth are so joined as to prevent the escape of gases into the atmosphere, as well as the escape of the material being dried into the flue below. The manner of providing for expansion and contraction of the hearth is simple and effective, consisting in anchoring each plate by means of lugs, *N*, which enter corresponding sockets im-

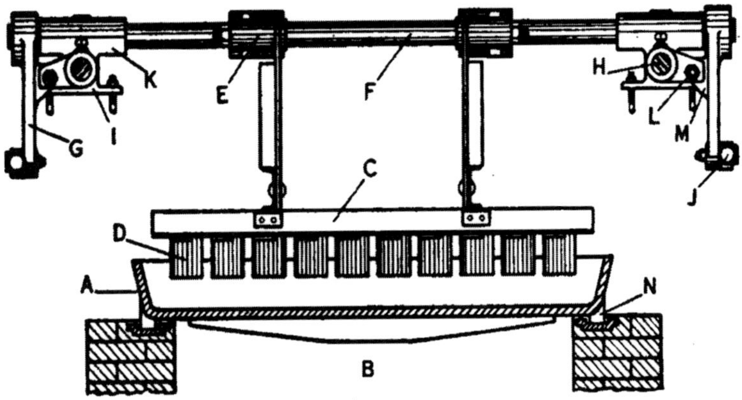

Fig. 185.

bedded in the side walls of the flue. Expansion of the hearth plates lengthwise is accommodated by the overlapping edges which have ample allowance for movement, and sidewise by the lugs *N*, sliding in their sockets which are designed to permit sufficient movement in that direction.

"The rake *C*, is built of structural steel, the transverse members bearing rabbles *D*, which on adjacent transverse bars are alternately right-hand and left-hand, for the purpose of imparting a plowing action to the material operated upon. These rabbles are adjustably attached to the transverse bars, permitting a wide variation of inclination, and consequently of speed of travel of the material over the hearth. This adjustability facilitates a rapid spreading of the feed over the entire width of the hearth where it is delivered at one point, or where receiving the discharge directly from a drum filter wider than the hearth, there is an excess at the two sides. The rabbles on succeeding transverse bars are so arranged that there is no tendency toward crowding the material to the side.

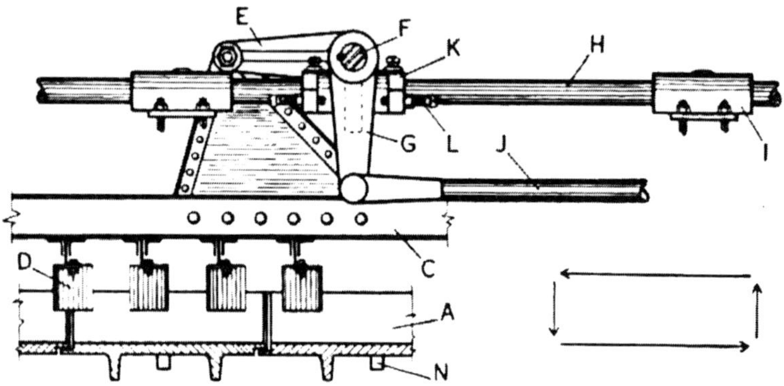

Fig. 186.

"The rake is supported and operated by the arms *E*, attached to the rock shafts *F*, which are journaled in carriages *K*, fixed to reciprocating bars *H*, free to move longitudinally in the bearings *I*. The arms *E* and *G*, together with the rock shaft *F*, form a bell crank, and the oscillation of the bell crank is limited by a lug *M*, on the arm *G*, and two lugs on the carriage *K*, the latter having set screws as the actual points engage to provide adjustability. These set screws, indicated at *L*, form the means of easy regulation. The set screws at the right in Fig. 186 control the depth to which the rake is lowered, and should be so adjusted that the rake just clears the bottom of the hearth. The one at the left controls the height to which the rake is lifted, but it is actually used to regulate the length of stroke, as increasing the lift reduces the stroke and contrariwise.

"The range of adjustment of stroke is from an amount less than the distance between the transverse rows of rabbles, to an amount exceeding it, so that the rabbles of one transverse series can be made to enter the furrows left by the adjacent ones, or what is usually better, to enter the ridges of material between the furrows, more effectually to chop it up and expose fresh surfaces. Sufficient lift is provided at all points of adjustment, and any lift in excess of the necessary amount is not great, and from an operating standpoint immaterial, as the weight of the rake is counterbalanced and practically no power is consumed. The counterbalancing weights have been omitted from the figures for the sake of clearness.

"The driving mechanism, not shown, consists of a shaft having a crank at each end, rotated slowly through the medium of suitable gearing. The main driving rods *J*, connect these cranks with the arms *G*, and rotation of the cranks would cause simple oscillation of the arms *G*, were it not for the stops on the carriages. These stops limit the oscillation *G* and *E*, and the excess of reciprocating motion causes a sliding of the bars *H*, in the bearings *I*.

"Fig. 186 shows the parts near the discharge end of the machine. This is also the driving end, as it is advantageous to pull rather than to push a rake of this kind, for the reason that all the parts are then in tension and vibration is avoided.

"The rake at the feed end is carried by a rock shaft and arms similar to those shown, and the lower ends of the corresponding vertical levers are connected to *G*, by side rods, not shown. However, there is this difference between the two supports and the two ends; at the discharge end a triangular rigid connection is used to receive the pin which works in the end of the arm *E*, whereas at the feed end a simple hanger, pin-connected at top and bottom, is provided, as the function of the rock shaft and arms at the end is only to lift and lower the rake. This flexibility of the feed and hangers permits unequal expansion of the rake and reciprocating bars. A further difference is that the carriages *K*, are provided with stops at the discharge end only, the side connecting rods causing the arms at the feed end to oscillate in unison with these at the discharge end. This removes the necessity of adjusting stops to act in unison with the stops at the discharge end, and permits free expansion from unequal heating without setting up strains.

"The travel of the rake is represented by the four arrows arranged in the form of a rectangle in the lower right-hand corner of Fig. 186. The horizontal movement, both forward and backward, is a truly straight one, and the vertical movement nearly so. This true horizontal movement is of the utmost importance from an efficiency standpoint, as it enables the rake to operate close to the hearth at all points, and prevents an accumulation of material at rest, which would act as an insulator of heat.

"From the foregoing description the operation will readily be seen to be as follows: Assuming the driving rod *J*, to be moving toward the right, the rake will be lowered till the lug on the arm *G*, engages the set screw stop *L*, when the reciprocating bar *H*, will slide to the right, carrying the rock shaft and rake with it, and this will continue till the crank passes the center and begins to move *J* to the left. As soon as the arm *J*, commences the movement toward the left, the contact of the lug on the arms *G*, with the set screw *L*, is broken, and the arm *G*, swings to the left until the lug engages the set screw on the left end of the carriage, this operation lifting the rake clear of the material on the hearth. Further movement of the arm *K*, pushes the rake backward until the driving crank passes the center, when the lowering preparatory to the forward stroke commences. This completes the cycle." (Colo. Iron Works Co., Denver Colo.)

(182) The efficiency which may be expected from such a dryer varies from 30 to 60 per cent., and the following illustrates the method of calculation:

QUANTITIES PER HOUR.

Feed, wet weight........	3,638 lbs.	Fuel Col. lignite.
Feed, dry weight........	2,583 lbs.	Calorific power.
Moisture in feed 29%...	1,005 lbs.	10,000 B. t. u. per lb.
Moisture in product 13%	382 lbs.	Fuel burned 166 lbs.
Moisture evaporated 16%	673 lbs.	Fuel per ton of feed 91.3
Water evaporated per lb. of fuel	4.05 lbs.	Fuel per ton of product 111.9.
Specific heat of solids assumed	0.25	
Temperature of feed....	52° F.	
Temperature of product.	212° F.	

CALCULATION.

Heat usefully applied:

In raising temperature of solids 2,583 × 0.25 × 160 =	103,320	B. t. u.
In raising temperature of total water 1,055 × 160 =	168,800	B. t. u.
In evaporating water 673 × 976 . . . . =	650,118	B. t. u.
Total heat usefully applied	922,238	B. t. u.
Total heat developed 166 × 10,000 . . . . =	1,666,000	

Whence efficiency = 55.5%.

PERFORMANCE OF DIFFERENT TYPES OF DRYERS.
(W. B. RUGGLES)

Type of drier	Double shell direct heat.	Indirect heat 705 sq. ft.	Single shell bricked-in direct heat.	Single shell direct heat.	Stationary with paddles direct heat.
Material	Sand	Coal	Cement slurry	Lime-stone	Nitrate of soda
Moisture, initial, per cent.	4.58	10.2	61.2	3 6	7.2
Moisture, final, per cent.	0.0	0.0	40.7	0.5	0.3
Calorific value of fuel B. t. u.	12100.0	12290.0	13200.0	13180.0	13600.0
Fuel consumed per hour, lbs.	398.0	213.6	667.0	460.0	87.0
Water evaporated per hour, lbs.	2196.0	924.2	4057.0	13 5.0	349.0
Water évap. per pound fuel, lbs.	5.3	4.3	6.1	2 3	4.0
Material dried per hour, lbs.	36460.0	8300.0	7680.0	41400.0	4581.0
Fuel per ton dried material, lbs.	21.8	51.3	17.3	22.2	38.0
Heat lost in exhaust air, per cent.	11.3	42.8	38.4	38.2	40.7
Heat lost by radiation, etc., per cent. . .	7.6	7.7	12.5	15.6	13.8
Heat used to evaporate water, per cent.	52.5	39.4	52.0	24.4	33.1
Heat used to raise temperature of material, per cent.	28.6	10.1	7.1	21.8	12.4
Total efficiency, per cent.	81.1	49.5	59.1	46.2	45.5

PERFORMANCE OF A STEAM DRYER.

Material: Starch feed. Moisture: initial 39.8%, final 0.22%. Dried material per hour 831 lbs. Water evaporated per pound steam, 0.691 lb. Temperature of material: moist 58°, dry 212°. Steam pressure, 98 lbs. gauge.

Total heat to evaporate 548 lbs. water at 58° into steam,

$548 \times (154.2 + 969.7) = 615{,}897$ B. t. u.

Heat supplied by 793 lbs. steam condenser to water at 212°,

$793 \times (1{,}188.2 - 180.3) = 799{,}265$ B. t. u.

Heat used to evaporate water,

$(615{,}897 \div 799{,}263) = 77.1\%$.

Heat used to raise temperature of material,

$(831 \times 154 \times 0.492) = 62{,}963 = 7.9\%$.

Loss by radiation: $100 - (77.1 + 7.9) = 15\%$.

Total efficiency: 85.0%.

WATER EVAPORATED AND HEAT REQUIRED FOR DRYING.

M = percentage of moisture in material to be dried.
Q = lbs. of water evaporated per ton (2,000 lbs.) of dry material.
H = British thermal units required for drying per ton of dry material.

M	Q	H	M	Q	H	M	Q	H
1	20.2	85,624	14	325.6	424,884	35	1,077	1,269,240
2	40.8	108,696	15	352.9	458,248	40	1,333	1,555,960
3	61.9	130.424	16	381.0	480,720	45	1,636	1,895,320
4	83.3	156,296	17	409.6	521,752	50	2,000	2,303,000
5	105.3	180,936	18	439.0	554,680	55	2,444	2,800,380
6	127.7	206,024	19	469.1	588,392	60	3,000	3.423,000
7	150.5	231,560	20	500.0	623,000	65	3.714	4,222,680
8	173.9	257,768	21	531.6	658,392	70	4,667	5,290,040
9	197.8	284,536	22	564.1	694,792	75	6,000	6,783,000
10	222.2	311,864	23	597.4	732,068	80	8,000	9,023,000
11	247.2	339,864	24	631.6	770,392	85	11,333	12,755,960
12	272.7	368,424	25	666.7	809,704	90	18,000	20 223,000
13	298.9	397,768	30	857.0	1,022,840	95	38,000	42,623,000

$$\text{Formulae; } Q = \frac{2000M}{100\text{-}M}\text{; } H = 1120Q + 63{,}000.$$

The value of H is found on the assumption that the moisture is heated from 62° to 212° and evaporated at that temperature, and that the specific heat of the material is 1.21.

$$2{,}000 \times (212 - 62) \times 0.21 = 63{,}000.$$

DRYERS AND DRYING.[1]

(183) A dryer is an apparatus for driving off the moisture from materials by the application of heat or by the absorption of the moisture in the material by means of mechancially induced currents. Due to the different physical and chemical properties of the various materials one and the same type of machine will not work satisfactorily on all kinds of material and care must be taken that the proper kind of drier is selected for the individual case under consideration.

The data required to make the selection is as follows:

First—The amount of free moisture in the material as it will be fed to the dryer.

Second—The amount of free moisture allowable in the material as it will be discharged from the dryer.

[1] Written especially for this book by F. W. Finch of the Ruggles-Coles Eng. Co.

Third—The temperature that the material may be raised to without its undergoing any undesirable change or on account of the danger of ignition of the material.

Fourth—Whether the products of combustion can come in contact with the material without injuring it.

Fifth—The action of the material as the moisture is driven off. This last can be described as the physical characteristics such as would be the case with a mixture of carbonate of lime and carbonate magnesia. This mixture will not only adhere readily to steel when drying, but in a rotary dryer tends to ball up the outer coating, being dry and acting as an insulation, thus preventing the inside of the ball from being heated and the moisture driven off. Nitrate of ammonia and nitrate of soda both tend to adhere to steel plates and build up in a hard mass which is difficult to break off. In cases such as the last, care must be exercised in selecting a machine that the design will take care of these difficulties.

The general classes of machines are:

"A" Direct heat dryers in which the gases of combustion come in contact with the material. These can be sub-divided into those dryers which will raise the material to a high temperature and those dryers which by means of special design, do not raise the material to a high temperature.

"B" Indirect heat dryers in which the gases of combustion do not come in contact with the material.

"C" Steam dryers in which the material or the air, according to design, is heated by means of steam pipes or coils; the moisture being absorbed by air currents. There are many designs and types of this class of machine.

"D" Vacuum dryers—both pan and rotary types.

CHAPTER XII.

DISTILLATION.

(184) The most familiar example of distillation is the preparation of whiskey, brandy or alcohol from fermented liquors. In the laboratory this is accomplished by means of dephlegmators or distillation tubes. One of the simplest of these consists of a wide glass tube filled with marbles or large glass beads.

(185) The object of the distillation is the more or less complete separation of water from alcohol. Water boils at 100° and alcohol at 80°. If we boil a mixture of 5 per cent. alcohol and 95 per cent. water the vapor at first given off will contain more than 5 per cent. alcohol and less than 95 per cent. water. As the mixed vapors ascend the tube they are condensed in part, the water condensing most easily. The vapor therefore grows richer in alcohol and poorer in water as it rises in the tube. If the tube were long enough we might reasonably expect that alcohol entirely free from water would issue from the side tubes at the top. We find, however, that no matter how long the tube, the alcohol will still contain about 5 per cent. water, so that distilled alcohol or cologne spirits contain 95 per cent. alcohol.

(186) Where there is no great tendency to combine, as is the case with alcohol and water, the separation may be more complete, but in no case is the separation absolutely perfect. In many cases, as in petroleum distillation, a partial separation is sufficient. Fig. 187 illustrates an apparatus for distilling acetic acid, made by Doulton and Co. In this case the dephlegmating column, Woulf bottles and condenser are made of stoneware.

In Figs. 188 and 189 two forms of rectifying still of copper made by the Walter E. Lummis Co., are illustrated.

(187) For Distilling Acids the Hart apparatus, shown in Fig. 190 is best adapted. This consists of a porcelain body *A* with hanging tubes of glass, open at the top and closed at the bottom, *B*. The acid is admitted at *E* and overflows at *G*. For hydrochloric acid the packing between tube and manifold may be of rubber, which must be kept from direct contact with the flame by an asbestos shield, but for nitric acid only asbestos fiber of the best quality carefully packed by an experienced hand will answer. This apparatus is defective in that there is no circulation in the tubes and, in distilling large quantities of crude acid

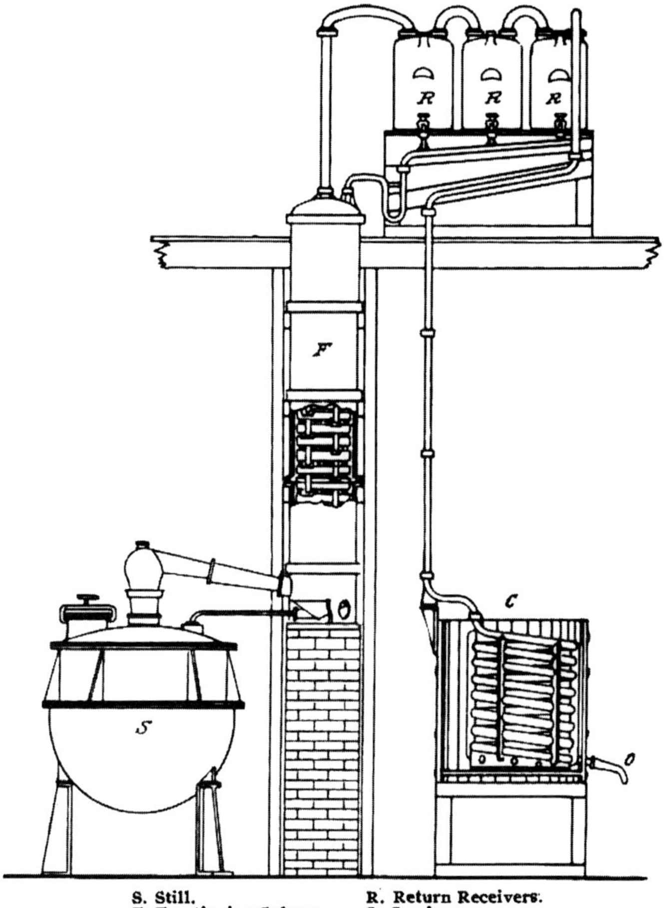

S. Still. R. Return Receivers.
F. Fractioning Column. C. Condenser.
O. Outlet for Pure Acid.

Fig. 187.—Column Apparatus For Distilling Acetic Acids.

containing dissolved salts, these are apt to deposit in the tubes. When the tubes become clogged the apparatus must be taken apart and cleaned.

The Walter E. Lummus Co., 173 Milk St., Boston, Mass.

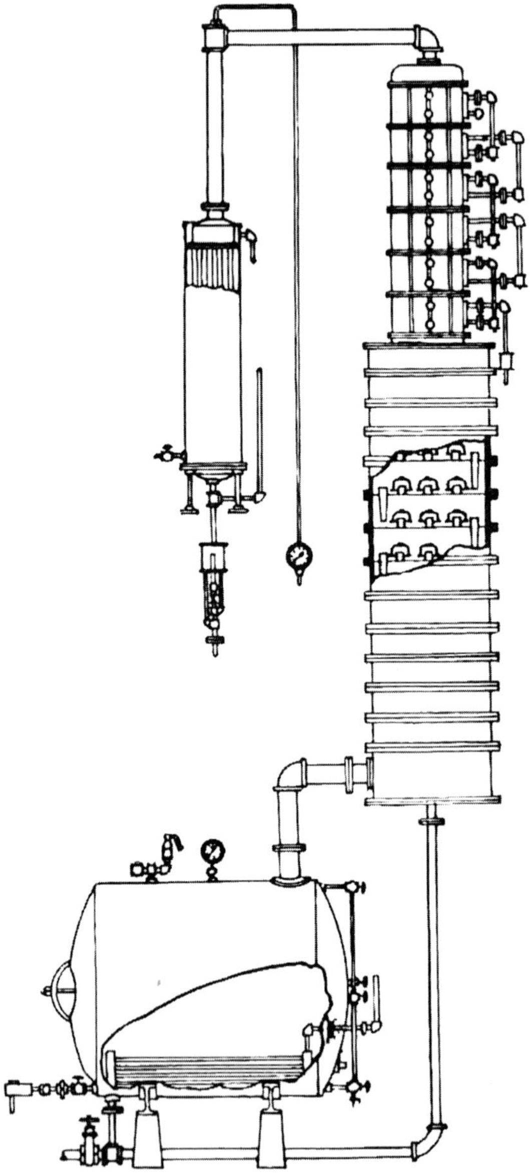

Fig. 188.—Benzol & Toluol Rectifying Still with Internal Reflux.

The Walter E. Lummus Co., 173 Milk St., Boston, Mass.

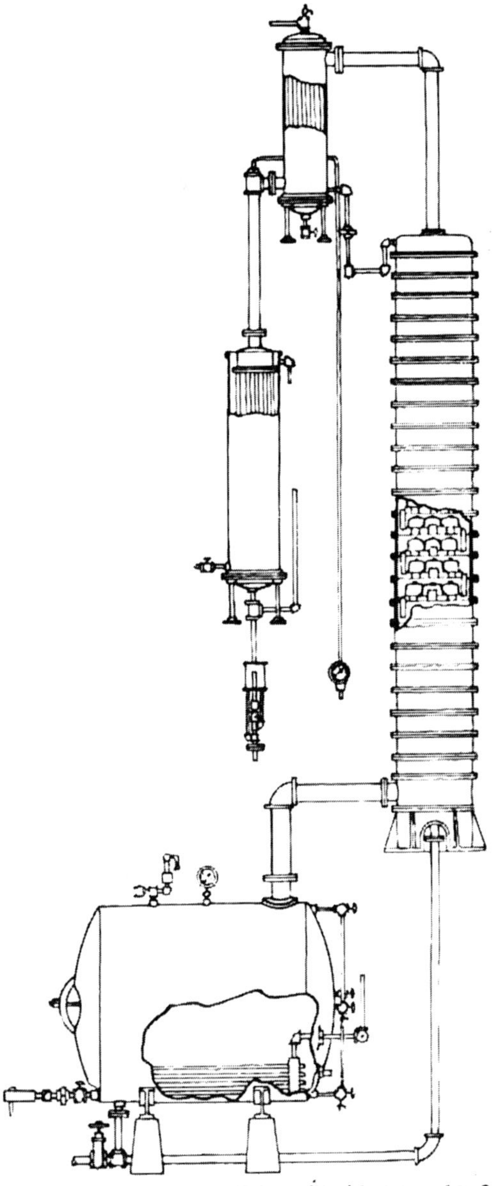

Fig. 189.—Benzol & Toluol Rectifying Still with External Reflux.

(188) For acid liquids containing at the same time dissolved salts and free acid, either the apparatus shown in Fig. 3 or the modified Hart boiler shown in Fig. 191 may be used. In the apparatus shown in Fig. 191, *E* is a leaden trough with leaden side tubes *J*, terminating over thistle tubes *F*, reaching to the bottom of the boiling tubes *B*. The evaporated liquid passes over at *G*, and the acid vapor goes off at *F*. Both this furnace and that shown in Fig. 190 are best fired with oil or gas. This form of

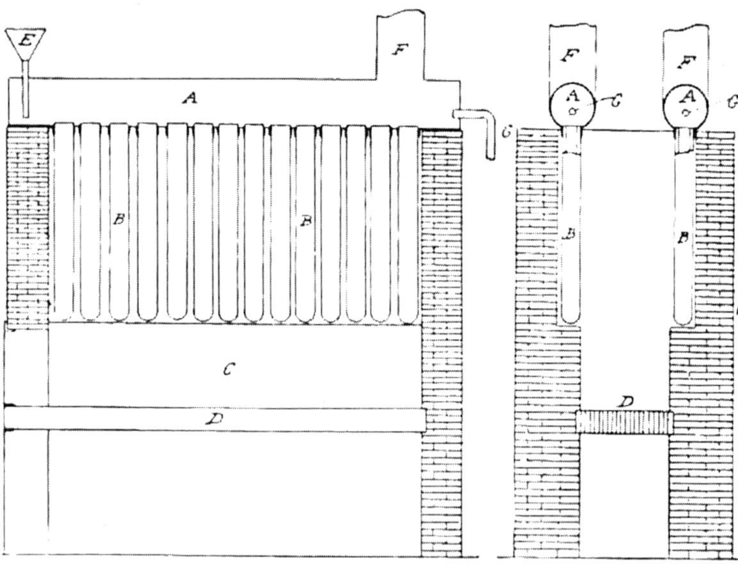

A. Acid receivers. D. Grates.
B. Acid tubes. E. Acid Inlet Funnel.
C. Furnace. F. Acid vapor exit pipes.
G. Residue outlet.

Fig. 190.—Apparatus For Distilling Acids.

apparatus though appearing fragile is very practical and has distilled many thousand tons of nitric and hydrochloric acids with great economy.

(189) Many Forms of Condenser have been used for the condensation of acid liquids. The simplest of these depends on air cooling for condensation, but this has generally been given up as too slow and uncertain in operation. A water-cooled worm is shown in Fig. 187. The Hart condenser for nitric acid is shown in Fig. 192. As here shown this is intended to produce a strong

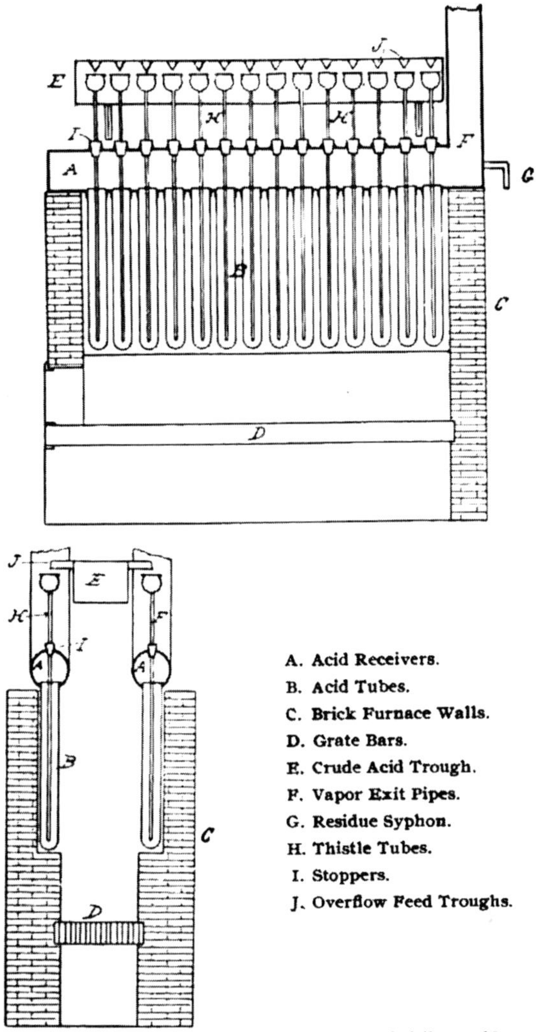

Fig. 191—Improved Apparatus—For Distilling Acids.

acid nearly free from nitrous acid. The vapor enters at *D* into the first manifold *C*, and then by the 3-inch glass tubes *B* into the second manifold. There are 12 of these tubes shown. At *J* the second manifold has a partition so that the remaining vapor must pass through the four tubes shown into the first manifold which also has a partition *J*. The remaining uncondensed vapor passes

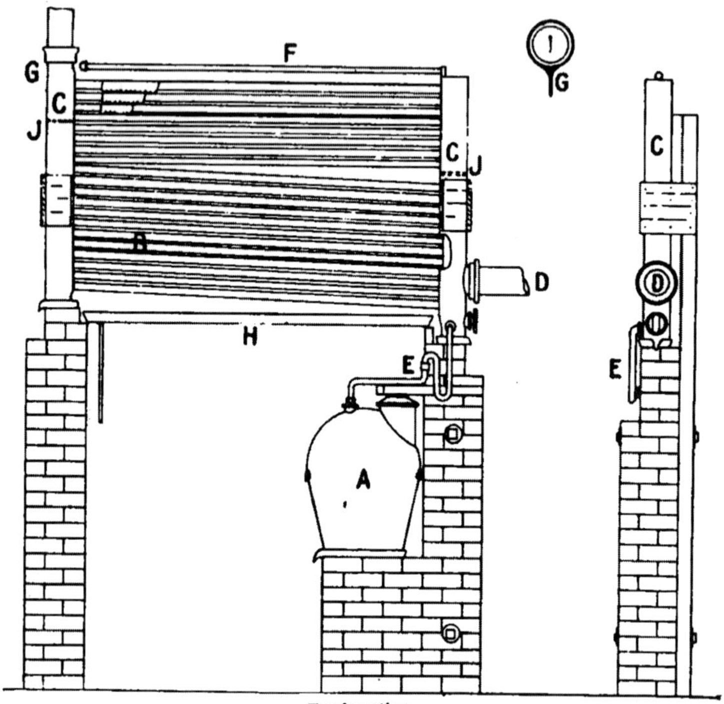

Explanation.

A. Receiver for HNO_3.
B. Condensing Tubes.
C. Stand Pipes.
D. Inlet Pipe.
E. Siphon.
F. Water Sprinkler.
G. Muslin Cover.
H. Waste Water Trough.
I. Showing manner of using muslin condense cover.
J. Stops.

Fig. 192.—Hart's Nitric Acid Condenser.

to the left through four tubes and to another condenser or to the towers or into the outer air by the tube *G*. The condensed acid runs backward in the tubes thus effecting a partial fractionation and the hot condensate collects in *A*. Water from a perforated pipe flows over the tubes and is distributed by means of muslin covers shown on the upper left hand corner and in section at *I*. This apparatus has the great advantage of taking little space, may be placed out of doors in freezing weather without loss of the tubes through freezing. If proper care is exercised it is very economical. Considerably more than half of the nitric acid made in the United States during the late war and much of that made abroad was condensed in this apparatus.

(190) Hough has devised a condenser in which the cooling coils are made of duriron. This answers for nitric acid, but has the disadvantage that if allowed to stand in the open in freezing weather it will be destroyed. The Moore and Hall condenser (U. S. Patent No. 986,846, Mar. 14, 1911) consists of an upright cylinder in which a number of very long test tubes are suspended. Into each of these tubes a stream of water is led, the nitric acid condensing on the outside. This also has the objection that it cannot be exposed to frost. The S condenser is built of large silica pipe rising above the retort in a series of flat curves, so that the condensed acid may run backward. This is a good condenser but is quite expensive.

14

CHAPTER XIV.

ABSORPTION OF GASES.

(191) Hydrochloric and hydrofloric acids and ammonia are usually sold as solutions in water. Solutions of carbon dioxide in water under pressure are sold as soda water. The solution of

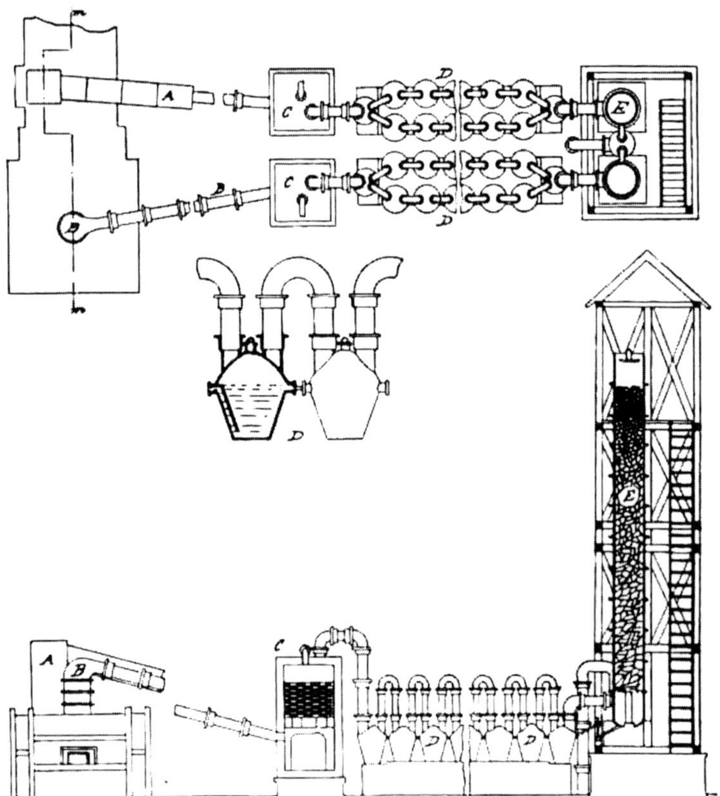

Fig. 193.—Hydrochloric Acid Condenser.

A. Roaster Gas. (Weak)
B. Pan Gas (Strong)
C. Cooling and Washing Tower.
D. Bombonnes or Absorbers. (30)
E. Coke Tower. (30 ft. high)
F. Fire Boxes.
P. Sulphate Pan.
R. Sulphate Roaster.
S. Opening for pushing pan sulphate cake into roaster.

several gases in water (SO_2 and NO_2) is used as a step in the manufacture of other substances (sulphites and bisulphites, nitric acid). In many other cases the solution is accompanied by evolution of considerable heat and as gases are most soluble in cold water means must be employed to keep the liquid at a temperature at least as low as that of the air if concentrated solutions are to be obtained.

Pützer—for Hydrochloric Acid

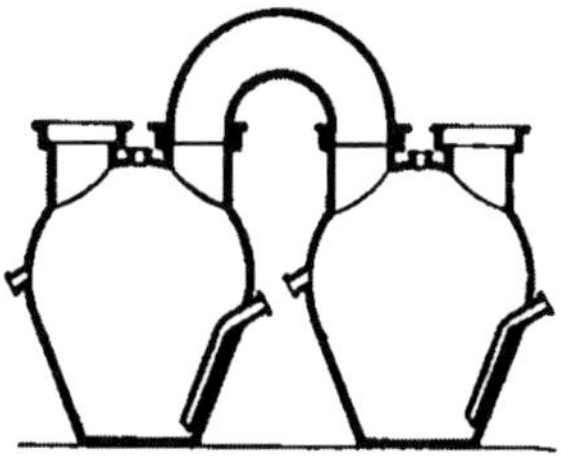

	Capacity gallons	Outside diameter inches	Height inches
I	43½	26¾	31½
II	52¾	29½	37½
III	59½	29½	34½
IV	66	31½	36¼
V	79¼	32¾	40¼
VI	118¾	35½	43¼

Normal—for Nitric Acid.

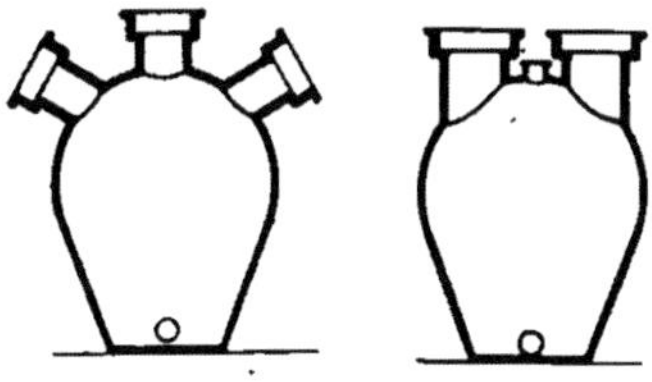

	Capacity gallons	Outside diameter inches	Height inches
I	14½	19¾	21¼
II	26½	21¾	28¼
III	39½	26	30
IV	43½	26¾	31½
V	52¾	29¼	37¾
VI	58	29½	34¾
VII	66	31½	36¼
VIII	79¼	32¾	40¼
IX	118¾	35½	53¼

Cellarius—for Nitric Acid.

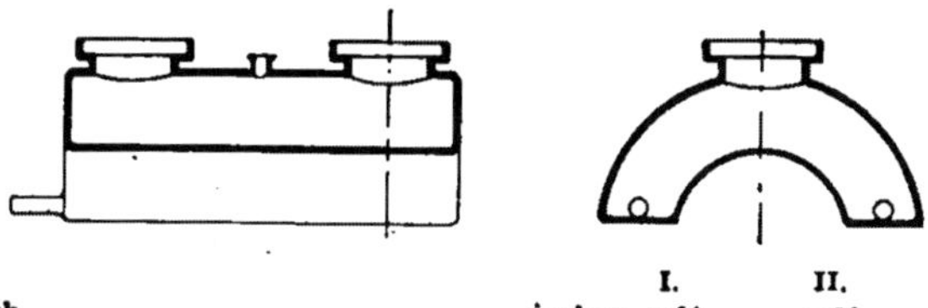

		I.	II.	III.
Length	inches	31½	39½	40¼
Width	"	19¾	26	35½
Height	"	12¼	15	20½
Approximate Cooling Surface	Sq. Ft.	9¼	15	22½
Gas Inlet and Outlet, Diameter	Inches	6	8¼	12¼
Liquid Inlet and Outlet	"	1½	2	2½
Capacity	Gallons	18½	29	66

Fig. 194.—Stoneware Tourills.

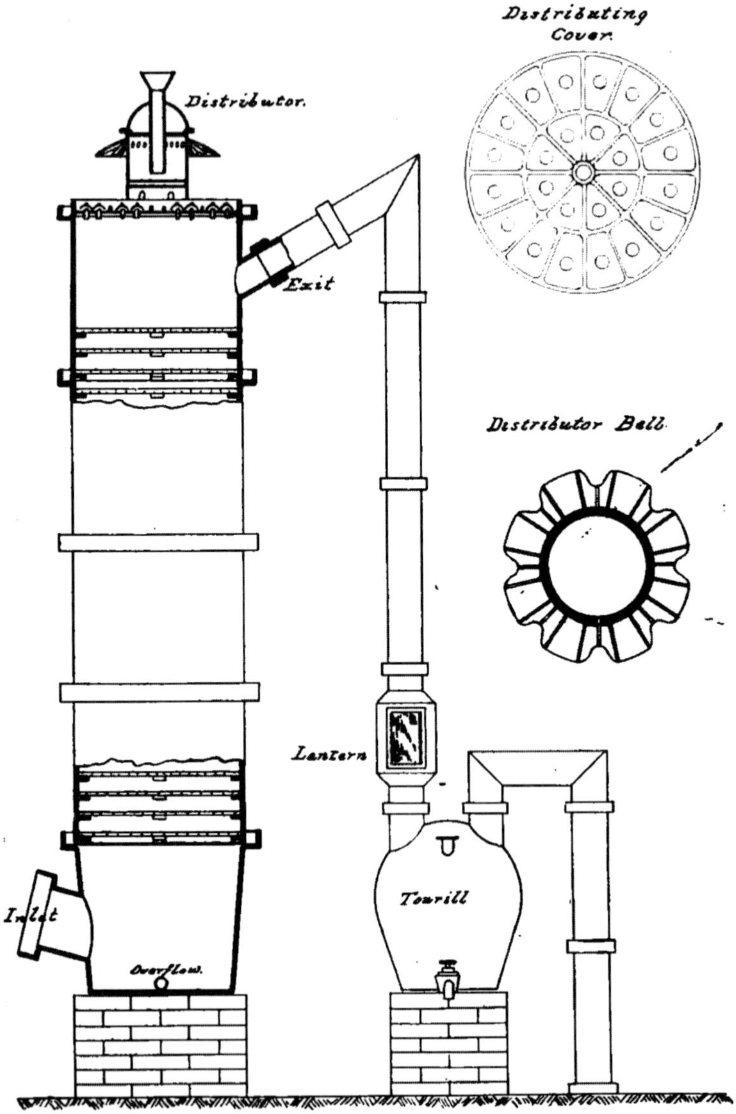

Fig. 195.—Circular Stoneware Towers. Lunge-Rohrmann-System.

(192) In the oldest form of apparatus, shown in Fig. 193, the gas, first properly cooled in a tower *C*, was passed through a series of Woulffe bottles connected at the tops by large inverted U

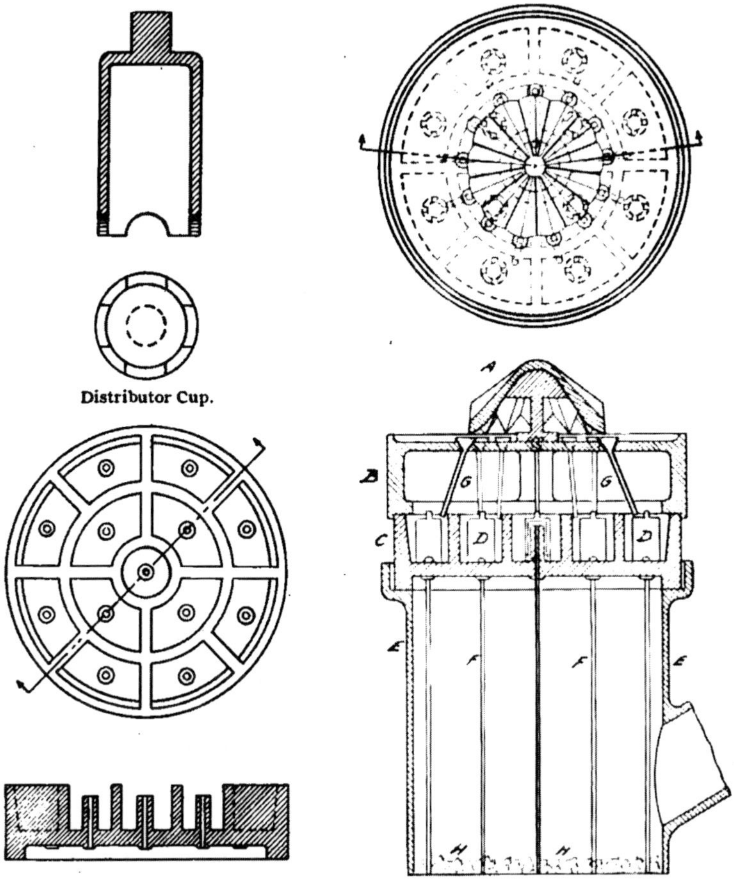

Distributor Plate. Distributor on Tower Top.

A. Simplex Distributor.
B. Funnel Plate.
C. Distributor Plate.
D. Distributor Cup.
E. Tower Top.
F. Hard Rubber Tubes.
G. Hard Rubber Funnels.
H. Packing.

Fig. 196.—Hydrochloric Acid Plant.

tubes. The solution obtained flowed from one bottle to another by side tubes as shown at *D*. The heat generated by solution was dissipated by air cooling or sometimes by setting the bottles in water. Other forms of stoneware bottle are shown in Fig. 194. This tower was faulty theoretically and did not lend itself to rapid production. The tower made of stoneware shown in Fig. 195 with details on 196, or of lead lined with chemical brick (Fig. 193) gave larger production but had the same theoretical defect and only relatively weak solutions were produced.

(193) In the Hart-Adamson absorber and cooler shown in Fig. 197, the gases pass in at *A* into the manifold C from which

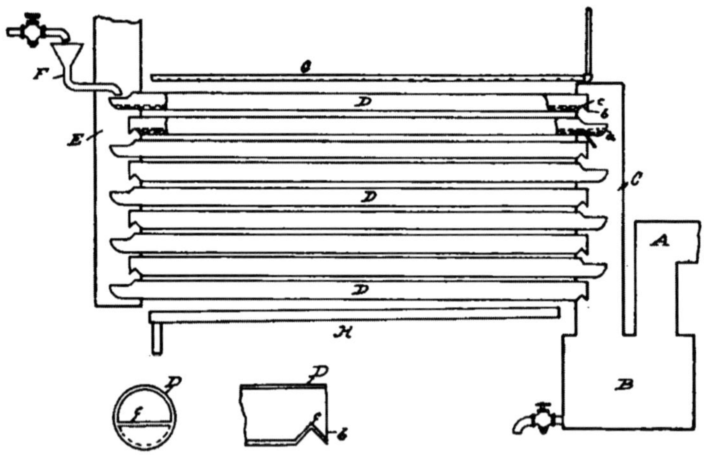

Fig. 197.—Acid Condenser. (Hydrochloric)

A. Acid Gas Inlet.
B. Acid Receiver.
C. Stand Pipe.
D. Condenser Pipes.
E. Stand Pipe.
F. Inlet for Absorbing Fluid.
G. Sprinkler.
H. Waste Water Trough.
a. Condenser Pipe Trough.
b. " " Spout.
c. Condenser Pipe Dam.

they pass through each of the tubes *D* into the manifold *E*. The absorbing liquid flows from the tap through the funnel *F*, through the first tube, back through the second tube, etc., until it reaches the vessel *B*, from which it is removed by the tap as desired. Each tube has a dam so that part of the liquid remains in the tube. Cooling is effected by running cold water over the tubes from the pipe *G*. The joints at either ends are made with rubber gaskets for hydrochloric acid. With this apparatus very strong fuming acids can readily be made in any quantity desired.

(194) The absorption of NO_2 and its conversion into reasonably strong HNO_3 offers considerable difficulty as the reactions occurring are of small velocity and consequently large surfaces are required. The heat liberation is negligible and so towers offer the cheapest solution. The theory of such towers is discussed by Partington and Parker in the *Journal of the Society of Chemical Industry,* Vol. XXXVIII, 1919, p. 75. Recently, another paper by Donnan and Masson on the theory of Gas Scrubbing Towers has appeared, *Ibid.,* **39,** 236, 1920.

CHAPTER XV.

MIXING AND KNEADING.

(195) In a great many operations mixing or kneading, according to the nature of the material must be accomplished. As illustrations of this we have the preparation of bread dough and the manufacture of spaghetti and macaroni, rubber mixing and kneading, the manufacture of graphite crucibles, crayon manufacture, dynamite, emery wheels, porcelain and pottery masses, etc., etc. For this purpose a great variety of machinery is manufactured. Among the prominent manufacturers of this class of machinery are the Werner & Pfleiderer Co., of Saginaw, Mich. Fig. 198 shows an asbestos mixer, Fig. 199 a vacuum masticator, and Fig. 200 shows how a battery of small experimental mixers may be driven from one countershaft.

Fig. 198.

Fig. 199.

Fig. 200.

CHAPTER XVI.

CONTAINERS.

(196) Containers may be of wood such as barrels and boxes, of fiber corrugated or plain and varnished or paraffined, of paper, metal or glass.

(197) **Barrels.**—A great variety of metal barrels are on the market. For small lots second hand sugar or flour barrels are often collected. Where large quantities of wooden barrels are used it is better to instal a barrel factory as part of the equipment. The capacity of a small barrel plant is from 250 barrels up. The staves and ends are bought in carload lots and three buildings will be needed, one for storing material, a second for the barrel making machinery and a third for barrel storage.

Barrel making machinery is made by the Peter Gerlach Co., Cleveland, Ohio, and the E. & B. Holmes Machinery Co., Buffalo, N. Y. Paper liners for barrels may be had.

(198) **Boxes** may be had in car-load lots put up in shooks with all the pieces of one box tacked together. Machines for driving a number of nails at one time so as to put such boxes together with a minimum of labor are on the market.

(199) **Corrugated Fiber Boxes** are sometimes satisfactory containers. Glass bottles and carboys are packed in box cars in salt hay or excelsior. Salt hay makes the best packing as it is very elastic. For acid carboys those with elastic slats or pieces of old rubber hose as resting surfaces are best.

(200) **Glass Packages** are usually charged at an advance on cost so as to cover breakage and cost of handling.

(201) **Paper or Cloth Bags** are sometimes used, as for cement. When possible, packages of all kinds should be non-returnable.

Index